Electronic and Electrical Servicing: Level 2

Consumer and Commercial Electronics core units

Electronic and Electrical Servicing: Level 2

Consumer and Commercial Electronics core units

Fourth Edition

Ian Sinclair

Geoff Lewis

E E B

ELSEVIER

AMSTERDAM · BOSTON · HEIDELBERG · LONDON · NEW YORK ·OXFORD
PARIS · SAN DIEGO · SAN FRANCISCO · SINGAPORE · SYDNEY · TOKYO

Newnes is an imprint of Elsevier

Newnes

Newnes is an imprint of Elsevier
Linacre House, Jordan Hill, Oxford OX2 8DP, UK
30 Corporate Drive, Suite 400, Burlington, MA 01803, USA

First edition 2002
Reprinted 2003, 2006

Notice
No responsibility is assumed by the publisher for any injury and/or damage to persons
or property as a matter of products liability, negligence or otherwise, or from any use
or operation of any methods, products, instructions or ideas contained in the material
herein. Because of rapid advances in the medical sciences, in particular, independent
verification of diagnoses and drug dosages should be made

British Library Cataloguing in Publication Data
A catalogue record for this book is available from the British Library

Library of Congress Cataloging-in-Publication Data
A catalog record for this book is available from the Library of Congress

ISBN–13: 978-0-7506-5423-4
ISBN–10: 0-7506-5423-6

For information on all Newnes publications
visit our website at www.newnespress.com

Printed and bound in *Malta*

06 07 08 09 10 10 9 8 7 6 5 4 3

Working together to grow
libraries in developing countries

www.elsevier.com | www.bookaid.org | www.sabre.org

ELSEVIER   BOOK AID International   Sabre Foundation

Contents

Preface to the fourth edition

This new edition continues the updating practice adopted in previous books in the series and reflects the rapid changes that are taking place within the electronics industry. The increase in the penetration of digital processing into the analogue world of home entertainment equipment continues to grow at such a pace that it does not need reference to a very large crystal ball to guess that the next edition will be almost completely based on digital concepts. In fact, since all signal processing is basically mathematical in form, the next breed of home entertainment systems are likely based on large scale digital processors with analogue to digital and digital to analogue converters as input and output interfaces.

Acknowledgements

The authors gratefully acknowledge the help provided by the City & Guilds of London Institute (CGLI), the Electronics Examination Board (EEB), and the Engineering & Marine Training Authority (EMTA) in the development of this series of books.

In addition, we would also like formally to recognize the contributions of the many lecturers and course tutors from the colleges of Further Education who have made many useful and constructive criticisms during the development of the previous three editions including *Electronics for the Service Engineer* and *Servicing Electronic Systems*.

Unit 1

1. Demonstrate an understanding of electrical units, primary cells and secondary cells and apply this knowledge safely in a practical situation.

2. Demonstrate an understanding of cables, connectors, lamps and fuses and apply this knowledge safely in a practical situation.

3. Demonstrate an understanding of resistors and potentiometers and apply this knowledge safely in a practical situation.

4. Demonstrate an understanding of health and safety in the training and work environment and apply this knowledge safely in a practical situation.

Note: Though health and safety appears as outcome 4 in the syllabus, we have started the book with this outcome. Safe working underlies all the later units, and we feel that it is better to concentrate all of these topics into the first chapter rather than deal with separate or overlapping health and safety topics in each other unit and outcome.

1

Health and safety

Health and safety considerations

The Health and Safety Commission (HSC) issues statistics on accidents at work, and the most recent are for the years 1999/2000. These reveal 187 255 reported accidents and a total of 1.2 million cases of musculo-skeletal disorders. These latter include cases ranging from back pain to repetitive strain injury. A worrying feature of these figures is that they are almost unchanged over several years.

The Health and Safety at Work Act of 1974 became fully operative in October 1978, and more recently the regulatory conditions have been further extended. This affects all aspects of servicing and other workshop activities. The legislation ensures that any work must be carried out in a way that ensures maximum safety for the employee and for anyone else present. Employers must provide a safe working environment and see that safety rules are obeyed. All employees (including students or apprentices) have a duty to observe the safety precautions that have been laid down by their employers, and to carry out all work in a safe way.

The 1974 Act can be summarized as follows:

All employers, including teachers, tutors and instructors, must ensure that any equipment to be used or serviced by employees, apprentices, or students is as safe to operate as it can reasonably be made. Equipment which cannot be made safe should be used only under close supervision.

Employers, tutors and instructors must make sure that employees and students at all times make use of the required safety equipment such as protective clothing, goggles, ear protectors etc. which the employer must provide for them.

Employees and students must ensure that they carry out all their work in such a way that it does not endanger them, people working around them, or the subsequent users of equipment that they repair.

Any accidents must be logged and reported, and measures taken to ensure that such incidents cannot happen again.

Failure to observe the provisions of the Act is reasonable grounds for dismissal.

Accident prevention depends on:

- recognition of hazards,
- elimination, if possible, of hazards,
- replacement of hazards,
- guarding and/or marking hazards,
- a sense of personal protection,
- continuing education in safety.

One particular difficulty that faces anyone working on servicing domestic electronics equipment is the wide variety of working places, which can vary from a well-designed workshop, as safe as possible in the light of current experience, to a home in which a TV receiver has failed, and in which almost every possible hazard, from bad electrical wiring to the presence of children around the TV receiver, exists. This huge difference in working conditions makes it vital for anyone working on electronics servicing to be aware of safe working methods and to practise them at all times whether supervised or not.

The types of hazards that lead to accidents can, as far as electronic servicing is concerned, are:

Electrical, particularly where industrial equipment is serviced.

Fire, particularly where flammable materials are used.

Asphyxia, particularly because of degreasing solvents.

Mechanical, particularly in cramped conditions.

Current health and safety legislation emphasizes accident *prevention* as well as dealing with accidents. Dealing with accidents is a matter of provision of facilities for, mainly, first aid and fire fighting and is very much in the hands of the employer. Prevention of accidents depends on attitudes of mind and is the responsibility of everyone concerned. The person who recognizes a hazard is the person who can imagine what might happen if an emergency arose, or who knows how to look for danger, and the Act implies that this person must not be treated as a trouble-maker or whistle-blower, but as a valuable contributor to safety who can save money in the long run.

The maximum penalty for failure to comply with the legislation is unlimited fines and up to two years' imprisonment. For the most serious cases, the charge of Corporate Manslaughter is now available.

The legislation now extends to the Control of Substances Harmful to Health (COSHH). In particular, this relates to cleaning solvents, paint, varnish, adhesives, sealants and even rosin-cored solder. Under these regulations a responsible person must be nominated to ensure that all related materials are stored and used safely. Such a person should also be familiar with the basic principles of chemical first aid procedures. With particular respect to the fumes emitted from rosin-cored solder, it is now well established that this can be the cause of asthma. Soldering should therefore ideally only be carried out within the confines of a fume extractor system. Recently developed rosin-free fluxes are now available, but even these should be used within a fume extractor environment. The cleaning solvents used for the removal of flux residues from printed circuit boards should not be chlorofluorocarbon (CFC) based as these also create environmental problems.

• The toxicity of lead is also recognized and as a result of this, tin/lead alloy solders are progressively being replaced by such as silver/copper alloy solders.

The regulations regarding the reporting of injuries and accidents at work have also been significantly reinforced. The Reporting of Injuries, Diseases and Dangerous Occurrences Regulations (RIDDOR) first appeared in 1986 but this has now been updated by RIDDOR95 which requires that employers and the self employed must notify the local environmental health department of any such occurrences.

Responsibility

The safety legislation requires both employer and employee to observe, maintain and improve safety standards and the adoption of safe working practices. The responsibilities of an employer are:

- To ensure that the workplace is structurally secure. Typical hazards are insecure floors, leaking roofs, blocked windows, restricted doorways and so on.

- To provide safe plant and equipment. Work benches and equipment must adequate for the job. All tools must be fitted with safety guards. Larger power tools should be screened or fenced. If necessary, floors should be marked with safe walking areas.

- To lay down safe working systems. Employees do not have to use makeshift equipment or methods. Protective clothing should be provided if needed.

- To ensure that environmental controls are used. The workshop must be kept at a reasonable working temperature. Humidity levels should be controlled if needed, and ventilation must be adequate. Employees should not be subjected to dust and fumes, and there must be washing facilities, sanitation for both men and women, and provision for first aid in the event of an accident.

- To ensure safety in the handling, storing and movement of goods (see below). Employees should not be required to lift heavy loads (a useful guide, enforced by law in some EU countries, is that an employee cannot be expected to lift without assistance a load of more than 15 kg). Mechanical handling must be provided where heavy loads are commonplace. Dangerous materials must be identified and stored where they do not cause any hazard to anyone on the site.

- To provide a system for logging and reporting accidents. Such a log is not to be simply a list of happenings, but a guide to better work practices.

- To provide information, training and updating of training in safety precautions, along with supervision that will ensure safety. This includes signs to indicate hazards.

- To devise and administer a safety policy that can be reviewed by representatives of both employer and employees.

Question 1.1

List the four main hazards of a servicing workshop.

Employees' responsibilities

The law recognizes that an employer can provide only a *framework* for safe working, and that it can be difficult to force an employee to work in a safe way. The Act therefore emphasizes that the employee also has responsibilities to ensure safe working. This is not just a matter of personal safety, but the safety of fellow workers and members of the public. Most accidents are caused by human carelessness in one form or another and though standards can be drawn up for safe methods of working, it is impossible to ensure that everyone will abide by these standards at all times.

The employees' responsibilities, which apply also to the self-employed, include:

Health-care. This means, for example, avoiding the use of alcohol or drugs if their effects would still be present in working hours. Remember that the term 'drugs' can include such items as painkillers, antihistamines for hay fever, and antidepressants. In some countries, blood alcohol level can be checked following industrial accidents as well as following a road accident. Working while excessively tired can also be a cause of accidents due to relaxed vigilance.

Personal tidiness. Clothing should provide reasonable protection, with no loose materials that can be caught in machinery or even (as has been reported) melted and set alight by a soldering iron. Long hair is even more dangerous in this respect, and must be fastened so that it cannot be caught.

Behaviour. Carelessness and recklessness cannot be tolerated, and practical jokes, no matter how traditional, do not belong in any modern workplace. Legislation in several countries now treats this type of behaviour in the workplace as seriously as it is treated on the roads.

Competence. An employee must know, either from experience, discussion with a colleague, or from reference to manuals, what has to be done and the safe way of carrying out the work. In several EU countries, it is an offence to carry out work for which you are not qualified, and if you are working abroad you will have to find out to what extent UK qualifications apply in other countries.

Deliberately avoiding hazards. Self-employed service personnel should insure against third party claims, and employees must remember that they can be sued if their careless or reckless behaviour leads to injury. Misusing safety equipment is a criminal offence, quite apart from endangering the lives of others. It is also an offence to fail to report a hazard which you have discovered.

Electrical safety

The main electrical hazard in servicing operations is working on live equipment, because of the ever-present risk of shock. Electric shock is caused by current flowing through the body, and it is the amount of current and where it flows that is important. The resistance of the body is not constant so that the amount of current that can flow for a given voltage will vary according to the moistness of the skin. The most important hazard is of electric current flowing through the heart. Ways of avoiding this include the following:

Ensure that only low voltages (less than 50 V) are present in a circuit.

Ensure that no currents exceeding 1 mA can flow in any circumstances.

Keep the hands dry at all times, because moist hands conduct much better than dry hands.

Keep workshop floors dry, and wear rubber soled shoes or boots.

Avoid two-handed actions, particularly if one hand can touch a circuit and the other hand is touching a metal chassis, metal bench or any other metal object.

The greatest hazard in most electronic servicing is the mains supply which in the UK and in most of Europe is at 240 V a.c. The correct use of three-pin plugs and sockets, correctly wired and fused, is essential. As a further precaution, earth leakage contact breakers (ELCBs) can be used to ensure that any small current through the earth line will operate a relay that will open the live connection, cutting off the supply.

If live working is unavoidable, try to avoid the possibility of current passing through the heart in the following ways:

- Work with the right hand only, making a longer electrical path to the heart in the event of touching live connections.
- Wear insulating gloves.
- Cover any metalwork which is earthed.

Power tools present a hazard in most workshops, and must be electrically safe. Some useful points are:

- Always disconnect when changing speed or drill bits.
- Never use a power tool unless the correct guards and other protective devices have been correctly fitted.
- Ensure that the mains lead is in perfect condition, with no fraying or kinking, and renew if necessary.
- Metal tools should be earthed, and in some conditions can be powered from isolating transformers. Another option is to use only battery powered tools.
- Most domestic power tools that employ plastic linings or inner casings to provide double insulation are not necessarily suitable for industrial applications.
- Test equipment must itself be electrically safe, using earthing or double insulation as appropriate.
- Test equipment can often be bought in battery-operated form.
- Mains powered test equipment must use the correct mains voltage setting, have power leads in good condition and the fusing correct for the load (usually 3 A).
- All test equipment, particularly if used on servicing live equipment, should be subject to safety checks at regular intervals.
- Users should maintain such equipment carefully, avoiding mechanical or electrical damage.

Low-power industrial equipment often makes use of standard domestic plugs, but where higher power electronic equipment is in use the plugs and socket will generally be of types designed for higher voltages and current, often for 3-phase 440 V a.c. In some countries, flat two-pin plugs and sockets are in use, with no earth provision except for cookers and washing machines, though by the end of the century uniform standards should prevail in Europe, certainly for new buildings.

Any mains electrical circuit should also include:

- A fuse or contact-breaker whose rating matches the maximum consumption of the equipment. This fuse may have blowing characteristics that differ from those of the fuse in the plug. It may, for example, be a fast-blowing type which will blow when submitted to a brief overload, or it may be of the slow-blow type that will withstand a mild overload for a period of several minutes.

- A double-pole switch that breaks both live and neutral lines. The earth line must *never* be broken by a switch.

- A mains warning light or indicator which is connected between the live and neutral lines.

All these items should be checked as part of any servicing operation, on a routine basis. In addition:

- As far as possible all testing should be done on equipment that is disconnected and switched off.

- The absence of a pilot light or the fact that a switch is in the OFF position should never be relied on as a sign that it is safe to touch conductors.

- Mains powered equipment being repaired should be completely isolated by unplugging from the mains.

- If the equipment is permanently wired then the fuses in the supply line must be removed before the covers are taken from the equipment.

Many pieces of industrial electronics equipment have safety switches built into the covers so that the mains supply is switched off at more than one point when the covers are removed.

The ideal to be aimed at is that only low-voltage battery-operated equipment should be operated on when live. When mains powered equipment must unavoidably be tested live, the meter, oscilloscope or other instrument(s) should not be connected until the equipment is switched off and isolated, and the live terminals should be covered before the supply voltage is restored. The supply line should be isolated again before the meter or other instrument clips are removed.

The main dangers of working on live circuits are:

- The risk of fatal shock through touching high voltage exposed terminals which can pass large currents through the body.

- The risk of fatal shock from the discharge of capacitors that were previously charged to a high voltage.

- The risk of damage to instruments or to the operator when a mild electric shock is experienced. The uncontrollable muscular jerking which is caused by an electric shock of any kind can cause the

Use a **DON'T SWITCH ON** sign if needed.

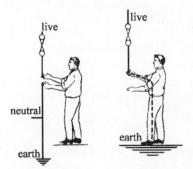

operator to drop meters, to lose his/her balance and fall on to other, possibly more dangerous, equipment.

Even low voltage circuits can be dangerous because of the high temperatures which can be momentarily generated when a short circuit occurs. These temperatures can cause burning, sometimes severe. As a precaution, chains, watchstraps and rings made of metal should not be worn while servicing is being carried out.

Older types of TV produced in the UK used 'live chassis' circuits in which the neutral lead of the mains supply was connected to the metal chassis of the equipment. Incorrect wiring of the plug or a disconnected neutral lead would cause the chassis to be live at full supply voltage. Later models of TV receiver used considerably lower internal voltages, but some still used a live chassis approach, and a few were wired so that the chassis was at about half of supply voltage, irrespective of how the plug was connected.

When carrying out service work either in the workshop or at a customer's premises, the equipment should only be powered via a mains isolation transformer. (see Chapter 6 for the operation of transformers.) This device has a 1:1 turns ratio with its magnetic core and metallic casing earthed to the mains supply input. The secondary winding is completely floating (isolated) so that there is no direct connection between the equipment being serviced and the mains power supply. For portable working, a transformer rated at about 150–250 watts is light enough to be readily portable, whilst a unit rated at about 500 watts is suitable for the workshop bench.

- Very high voltage levels at high currents can be found in microwave oven circuits, and these should not be switched on with the covers removed. Other white goods are likely to use mains voltages for heaters and motors, and there are the additional mechanical hazards from motors and drive belts.

Other electrical safety points are:

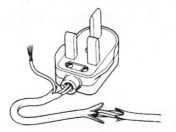

Cables should not be frayed, split, or be sharply bent, too tightly clamped, or cut.

Damaged cables must be renewed at once.

Hot soldering irons must be kept well away from cable insulation.

Cables must be securely fastened both into plugs and into powered equipment. The supply cable to a heavy piece of equipment should be connected by way of a plug and socket which will part if the cable is pulled.

The live end of the connector must have no pins which can be touched.

Every electrical joint should be mechanically as well as electrically sound. It must be well secured with no danger of working loose and with no stray pieces of wire.

The old adage of 'volts jolt but mills kill' implies that only a few milliamps of through-body current can be fatal.

- The old-fashioned type of lead/acid accumulator releases an explosive mixture of hydrogen and oxygen while it is on charge, and this can be ignited by a spark. No attempt should ever be made to connect or disconnect such accumulators from a charger or from a load while current is flowing. An explosion is serious enough, but the explosive spray of acid along with sharp glass or plastic fragments is even worse.

Question 1.2

List the five precautions that can be taken against electric shock.

Portable appliance testing (PAT)

It is an HSE regulation requirement that the flexible leads/cables, plugs and sockets of all electrical equipment should be examined and tested at regular intervals, typically at least annually and *always* after repairs. This operation should be carried out by a technically competent person. Once tested, each appliance should be suitably labelled and a recording of the results kept in a readily accessible database.

The cables and connectors should be physically examined for serviceability, defects and to ensure that they meet the current technical standards. The plugs should have shrouded live and neutral pins, the cables must be correctly colour coded and no component should be damaged in any way. The cord grip screws should prevent the cable being twisted from beneath its clamp. If a line connector and socket is in use, then the socket must be connected nearest to the mains supply point to ensure that contact with the live mains is prevented. Similarly, the connections at the appliance end should be sound and securely clamped.

If there are externally accessible fuse holders, these should be undamaged and loaded with fuses of the correct rating. The mains supply on/off switch should be undamaged and work correctly. If the output is greater than 50 volts, then the short circuit current must be limited to less than 5 mA.

Both single and double insulation types of equipment have to be considered and in the latter case, the appliance is not directly earthed and the case should carry the double square symbol (one within the other). The earth bonding test which applies only to the former types of appliance consists of measuring the earth wire resistance whilst passing a current of up to 25 amps for up to 5 seconds. This ensures that any poor earthing connections will become open circuit under test. The total resistance including the earth line and the resistance to the casing should not exceed $0.75 \, \Omega$.

Insulation testing is carried out at a potential of at least 500 volts DC. For directly earthed appliances the line and neutral pins are shorted together and the resistance measured between them and the earth pin. A value of at least $2 \, M\Omega$ is considered to be satisfactory. For double insulated appliances, the resistance between the shorted pins and any exposed metal work should be measured where a value of more than 7 MΩ is acceptable.

PAT testing kits range from relatively low cost, hand held systems up to larger less portable devices that carry LCD screens and contain interfaces to a PC/printer for data logging.

Fire precautions

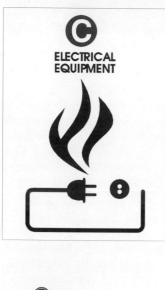

ELECTRICAL EQUIPMENT

Workshop fires present several types of hazard. The obvious one is flesh burns, but injuries from other causes are common, such as falling when running from a burning workshop. Clear escape routes should be marked and these must be kept clear at all times. If there is further risk of suffocation by smoke, low-level emergency lighting should be present so that a route to a safe exit is marked.

A fire can start wherever there is material that can burn (combustible material), air, or oxygen-rich chemicals, that can supply oxygen for burning, and any hot object that can raise the temperature of materials to the burning point. A fire can be extinguished by removing all combustible material, by removing the supply of air (oxygen), by smothering with non-combustible gas or foam, or by cooling the material to below the burning temperature. The most dangerous fires are those in which the burning material can supply its own oxygen (such materials are classed as *explosives*) and those in which the burning material is a liquid that can flow about the workplace taking the fire with it.

Another major danger is of asphyxiation from the fumes produced by the fire. In electronics workshops the materials that are used as switch cleaners, the wax in capacitors, the plastics casings, the insulation of transformers and the selenium that can still be found in some old-fashioned metal rectifiers will all produce dangerous fumes, either choking or toxic, when burning. Good ventilation can reduce much of this particular hazard.

The key points for fire safety are:

- Good maintenance
- No naked flames and no smoking permitted
- Tidy working, with no accumulation of rubbish
- Clearly marked escape paths in case of fire
- Good ventilation to reduce the build-up of dangerous fumes
- Suitable fire extinguishers in clearly marked positions
- Knowledge by all staff of how to deal with fire/explosion
- Regular fire drills and inspection of equipment.

Because of the variety of materials that can cause a fire, more than one method of extinguishing a fire may have to be used, and use of the wrong type of extinguisher can sometimes make a fire worse. The five main types of fire extinguisher and their colour coding are:

water based (**red**)   foam (**cream**)   powder (**blue**) CO_2 (carbon dioxide) gas (**black**)   inert liquid (**green**)

The water based type of extinguisher (for a *Class A* fire) is most effective on materials that absorb water and which will be cooled easily. Fires in paper, wood or cloth are best tackled in this way, using the extinguisher on the base of the fire to wet materials that are not yet burning and to cool materials that have caught light. Water based extinguishers must *never* be used on electrical fires or on fires that occur near to electrical equipment.

Class B fires involve burning liquids or materials that will melt to liquids when hot. The main hazards here are fierce flames as the heat vaporizes the liquid, and the ease with which the fire can spread so as to affect other materials. The most effective treatment is to remove the air supply by smothering the fire, and the foam, dry powder or CO_2 gas extinguishers can all be useful, though such extinguishers should be used so that the foam, powder or gas falls down on to the fire, because if you direct the extinguisher at the base of this type of fire the liquid will often simply float away, still burning. Fire blankets can be effective on small fires of this type, but on larger fires there is a risk that the blanket will simply act as a wick, encouraging the fire by allowing the burning liquid to spread.

For dealing with electrical fires, the water or foam based types should be avoided because the risk of electric shock caused by the conduction of the water or foam is often more serious than the effect of a small fire. Inert liquid extinguishers are effective on electrical fires but the liquids can generate toxic fumes and can also dissolve some insulating materials. Powder extinguishers and CO_2 types are very effective on small fires of the type that are likely to develop in electronics workshops. They must be inspected and check-weighed regularly to ensure that internal pressure is being maintained.

A few sand buckets are also desirable. They must be kept full of clean sand and never used as ashtrays or for waste materials. A firemat is also an important accessory in the event of setting fire to the clothing of anyone in the workshop. The firemat should be kept in a prominent place and everyone in the workshop should know how to use it.

In the event of a fire, your order of priorities should be:

- To raise the alarm and call the fire brigade even if you think you can tackle the fire.
- To try to ensure evacuation of the workshop.
- To make use of the appropriate fire extinguishers.

Question 1.3

For what types of fire is a water based extinguisher suitable? When must this type of extinguisher *never* be used?

The important point here is that you must never try to fight a fire alone, nor to put others at risk by failing to sound the alarm. The most frightening aspect of fire is the way that a small flame can in a few seconds become a massive conflagration, completely out of hand, and though prompt use of an extinguisher might stop the fire at an early stage, raising the alarm and calling for assistance is more important.

General safety
Soldering

Soldering should never be carried out on any equipment, whether the chassis is live or not, until the chassis has been completely disconnected from the mains supply and all capacitors safely discharged. The metal tip of the soldering iron will normally be earthed, and should not be allowed to come into contact with any metalwork that is connected to the neutral line of the mains, because large currents can flow between neutral and earth. Obviously, any contact between the soldering iron and the live supply should also be avoided.

Soldering and desoldering present hazards which are peculiar to the electronics workshop. The soldering iron should always be kept in a covered holder to prevent accidental contact with hands or cables. Many electrical fires are started by a soldering iron falling on to its own cable or the cable of another power tool or instrument. Though it can be very convenient to hang an iron up on a piece of metal, the use of a proper stand with a substantial heat sink is the safe method that ought to be used.

In use, excessive solder should be wiped from the iron with a damp cloth rather than by being flicked off and scattered around the workshop. During desoldering, drops of solder should not be allowed to drip from a joint; they should be gathered up on the iron and then wiped off it, or sucked up by a desoldering gun which is equipped with a solder pump.

Care should be taken to avoid breathing in the fumes from hot flux. Where soldering is carried out on a routine basis, an extractor hood should be used to ensure that fumes are efficiently removed. Materials other than soldering flux can cause fumes, and some types of plastics, particularly the PTFE types of materials and the vinyls, can give off very toxic fumes if they are heated to high temperatures. Extraction can help here, but it is preferable to use careful methods of working which ensure that these materials are not heated.

Toxic materials

Virtually every workplace contains toxic materials, or materials that can become toxic in some circumstances, such as a fire. Industrial solvents, such as are used for cleaning electronics subassemblies, switch contacts etc. are often capable of causing asphyxiation and even in small concentrations can cause drowsiness and stupor. Many common insulating materials are safe at normal temperatures, but can give off toxic fumes when hot.

In addition, some very poisonous materials are used within electronics components. Such materials can be found in cathode-ray tubes, fluorescent tubes, valves, magnetrons (used in microwave ovens), metal rectifiers, power transistors, electrolytic capacitors and other items. The local hospital should be informed of the toxic materials that are present, and, if possible, should advise on any first aid that might be effective in the event of these materials being released. In particular, cathode-ray tubes and fluorescent tubes must be handled with care because they present the hazard of flying glass as well as of toxic materials if they are shattered.

Transistors and ICs should never in any circumstances be cut open. No servicing operation would ever call for this to be done, but if a faulty component has to be cut away there might be a danger of puncturing it. Some power transistors, particularly those used on transmitters, contain the solid material beryllium oxide whose dust is extremely poisonous if inhaled, even in very small quantities. All such transistors that have to be replaced should be removed very carefully from their boards and returned to the manufacturers for safe disposal.

Corrosive chemicals form another class of toxic material which cause severe damage to the skin on contact. Sulphuric acid has the effect of removing water from the skin, causing severe burning, whereas sodium hydroxide (caustic soda) dissolves fatty materials and thus damages the skin by removal of fat. All strongly acidic or alkaline materials will cause severe skin damage on contact, but by far the most dangerous in addition to sulphuric acid and sodium hydroxide are hydrofluoric acid (used for etching glass) and nitric acid (used in etching copper). The only first aid treatment is to apply large amounts of water to dilute the corrosive material. Even very dilute acids or alkalis will cause severe damage if they reach the eye, and once again, large amounts of pure water should be applied as a first aid measure. *Never* attempt to neutralize one chemical with another, because neutralization is usually accompanied by the generation of heat. Specialized treatment at hospital should always be sought in any case of accident with corrosive materials.

Use of hand tools

The incorrect use of small hand tools is a very common cause of accidents and safe methods of working must always be used *especially* when you are in a hurry.

Never point a screwdriver at your hand. This seems obvious, yet a common cause of stabbing accidents is using a screwdriver to unscrew a wire from a plug that is held in the palm of the hand rather than between fingers or, better still, in a vice.

Small tools have to be suited to the job and used correctly. Never use a blade screwdriver when a Philips or Pozidrive type is needed, and remember that screwdriver size must be matched to the work.

Files should be fitted with handles and used with care because, being brittle, they can snap. Never use a file as a tommybar, for instance.

Box spanners and socket sets should be used in preference to open spanners where possible.

Snipping wire with sidecutters can be a hazard to the eyes, either of the user of the cutters or of anyone standing close. Ideally, the wire which is being snipped should be secured at both ends so that no loose bits can fly around. One safe method is to hold the main part of the wire in a vice and the other part in a Mole wrench.

Protective clothing should be worn wherever the regulations or plain common sense demands it. The workshop is no place for loose ties or cravats, for long untied hair or strings of beads. A workshop coat or boiler suit should be worn whenever workshop tools are being used, and eyeshields or goggles are nearly always necessary. Goggles and gloves must always be worn when cathode-ray tubes are being handled and safety boots should be worn when heavy objects have to be moved. Goggles are also a useful protection when wire is being snipped, particularly for hard wire like Nichrome.

Back injury

Back injury is the single most common cause of absence from work in the UK. In the course of electronics servicing work many large and heavy items have to be shifted. Injury is most commonly caused by incorrect methods of lifting heavy objects or holding such objects before putting them into place.

The correct method of lifting is to keep a straight back throughout the whole of the lifting operation, bending your knees as necessary but *never* your back.

Never attempt to shift anything heavy by yourself.

Even if you bear all the weight of carrying a load, help in lifting and steering the load can make the difference between safe and unsafe work.

In several countries, there is a statutory upper limit of load that one operator is allowed to lift without help. This can be as low as 15 kg (33 lb), which is less than the usual 20 kg (44 lb) weight allowance for luggage on flights.

Workshop benches and stools should be constructed so that excessive lifting is not required.

If a large number of heavy items have to be moved to and from benches, mechanical handling equipment should be used.

Stool heights should be adjustable so that no user needs to stoop for long periods to work on equipment.

Reporting hazards and accidents

An employer must ensure that a workspace is kept free of dangerous and badly sited materials. Someone in the workshop, however, may have put the hazard into place, and quite certainly someone who works near the hazard will be the first to notice it. The responsibility cannot be shrugged off as being entirely that of the employer.

Two logbooks should be maintained for any workshop. One is used for reporting potential hazards so that action can be taken (or if the worst happens, blame apportioned). The other logbook is used to record actual accidents, so that a written account is available of every incident.

Referring to these logs at regular intervals, perhaps at the start of each month, is a valuable way of revising safety policies.

First aid

In every workshop at least one person, and preferably two should be trained in first aid procedures. Particular attention should be given to the procedures for resuscitating victims of electric shock, but you should remember that most first aid cases will require attention to fairly minor cuts and bruises.

First aid need not involve elaborate treatment, and most first-aid is concerned with minor cuts and burns.

A fairly basic first aid chest along with good facilities for washing is adequate for at least 90 percent of incidents.

Treatments for exposure to toxic materials or fumes call for much more specialized equipment and knowledge.

It is particularly important never to work alone with potentially toxic materials.

Training to deal with electric shock should be a first priority of first aid instruction for servicing workshops.

The power supply must be cut off before the victim is touched, otherwise there may be two victims instead of one.

Mouth to mouth resuscitation, preferably by way of a plastic disposable mouthpiece, should then be applied as soon as the power is off, and continued until the victim is breathing or until an ambulance arrives.

If it would take too long to find and operate a mains switch, push the victim away from live wires using any insulated materials, such as a dry broom handle, or pull away by gripping (dry) clothing.

Do not touch the skin of the victim. This endangers the rescuer and also the victim (because of the additional current that will flow, perhaps through the heart).

Speed is important, because, even if breathing can be restored, the brain can be irreparably damaged after about four minutes.

Always leave an unconscious victim on his/her side, never on the back or on the face.

Never try to administer brandy or any other liquid to anyone who is not fully conscious because the risk of choking to death is at least as great as that of electric shock itself.

Once the victim has been removed from the danger of continued shock, a 999 call can be made and mouth to mouth resuscitation can be given. Practical experience in this work is essential, and should form an important part of any first aid training. A summary will remind you of the steps, and posters showing the method in use should be displayed in the workshop.

Place the victim on his/her back, loosen any clothing around the neck, remove any items from the mouth such as false teeth or chewing gum.

Tip the head back by putting one of your hands under the neck and the other on the forehead. This opens the breathing passages.

Pinch the nose to avoid air leakage, breathe in deeply, and blow the air out into the victim's lungs. If possible, use an approved mouth to mouth adapter to avoid any risk of transferring disease, but never waste time looking for one.

Release your mouth and watch the victim breathe out. You may have to assist by pressing on the chest.

Repeat at a slow breathing rate until help arrives or the victim can breathe unattended. Do not give up just because the victim is not breathing after a few minutes because these efforts can sustain life and avoid brain damage even if the victim is unconscious for hours.

Electric shock is often accompanied by the symptoms of burning which will also have to be treated, though not so urgently. The workshop telephone should have permanently and prominently placed next to it a list of the numbers of emergency services such as doctors on call, ambulance, fire, chemists, hospital casualty units, police, and any specialized services such as burns and shock units. This list should be typed or printed legibly, maintained up to date and stuck securely to a piece of plywood or hardboard. Part of any safety inspection should deal with checking that this list is updated, well-placed, and legible.

Severe burning must be treated quickly at a hospital. First aid can concentrate on cooling the burns and treating the patient for shock.

- Apply cold water to the region of the burn and when the skin has had time to cool, cover with a clean bandage or cloth.

- Never burst blisters or apply ointments, and do not attempt to remove burned clothing because this will often remove skin as well.

- The reaction to burning is often as important as the burn itself, and any rings, bracelets, tight belts and other tight items of wear should be removed in case of swelling.

- Try to keep the patient conscious, giving small drinks of cold water, until specialist help arrives.

- The treatment of minor wounds is a frequent cause of a call on first aid. The most important first step is to ensure that a wound is clean, washing in water if there is any dirt around or in the wound. Minor bleeding will often stop of its own accord, or a styptic pencil (alum stick) can be used to make the blood clot.

- More extensive bleeding must be treated by applying pressure and putting on a fairly tight sterile dressing one dressing can be put over the top of an older one if necessary rather than disturbing a wound. Medical help should always be summoned for severe wounding or loss of blood, because an anti-tetanus injection may be required even if the effects of the wound are not serious.

The effects of chemicals require specialized treatment, but first aid can assist considerably by reducing the exposure time. In electronics servicing work, the risk of swallowing poisonous substances is fairly small, and the main risks are of skin contact with corrosive or poisonous materials and the inhalation of poisonous fumes.

If a corrosive chemical has been swallowed or spilled on the skin, large amounts of water should be used (swallowed or used for washing) to ensure that the material is diluted to an extent that makes it less dangerous.

Common solvents like trichloroethylene and carbon tetrachloride degreasing liquids give off toxic fumes. These liquids should be used (other than in very small quantities) only under extractor hoods or in other well ventilated situations.

Never try to identify a solvent by sniffing at a bottle. Bottles should be correctly labelled, preferably with a hazard notice.

Some solvents, like acetone and amyl acetate, are a serious fire hazard in addition to giving off toxic fumes.

Care of equipment under repair

While equipment is being serviced, it is the responsibility of the servicing engineers to ensure that the equipment is not damaged. Damage in this respect means mechanical or electrical damage, and though such hazards should be covered by insurance, the customer is unlikely to feel well disposed to any servicing establishment which cannot take better care of equipment that is being serviced.

Mechanical damage covers such items as:

- Damage to either the cabinet or the functioning of the circuitry, or both, caused by dropping the equipment.
- Tool damage, such as burn marks on a cabinet caused by a soldering iron.
- Marks or stains, such as can be caused by a hot coffee cup or by carelessness with solvents.

Such mechanical damage might be due to an unavoidable accident, but only too often it is the result of carelessness, and the ultimate loser, regardless of insurance, will be the staff of the workshop. Never assume that an old piece of equipment is of low value to the customer, or that a customer will not worry about damage. Since mechanical damage to the cabinet is the form of damage that is most obvious to the customer, all cabinets should be wrapped both when being handled and in the course of servicing. Bubble-pack is particularly useful for avoiding damage caused by knocks, and this can also protect against marking or staining. Damage from soldering irons can be minimized by using holders for irons, and ensuring that cabinets are kept well away from tools.

One other hazard of this type is ingrained swarf, including fragments of solder caused by shaking a soldering iron. If the workshop is not kept clean, metal swarf can become embedded in the underside of a cabinet. Though this will not be visible to the customer, it can cause severe scratching on the table that the equipment stands on, and the customer is not likely to forget or forgive the damage.

Electrical damage is typically caused by:

Incorrect use of servicing equipment such as signal generators.

Carelessness in removing ICs.

Ignorance of correct operating conditions.

Replacement of defective components by unsuitable substitutes.

Operating equipment with incorrect loads (including o/c or s/c load).

Failure to check power supply components when the initial fault has been caused by a power-supply fault.

One feature that is common to many of these items is incorrect or unavailable information. Equipment should not be serviced if no service sheet is available. Granted that there are some items which are so standardized that no service sheet is necessary, but these items are the ones, such as radios and personal stereo players, that are uneconomical to service in any case. Equipment such as TV receivers, hi-fi, personal computers, video-recorders etc. will require information to be available, and that information should be as complete as possible. Service engineers must be aware which components must be replaced only by spares approved by the manufacturers, and why. Substitution should be considered only if the manufacturer no longer supplies spares – remember that some Japanese manufacturers will no longer supply spares after a comparatively short period, as short as 5 years.

Outside installation work

Outside installation work, as distinct from outside servicing, is unavoidable, and for consumer electronics often refers to aerials (FM, TV or satellite), down-leads and interconnections. The installation of personal computers, however, also comes into this class, and though aerial work is specialized and will usually be contracted out (and will not be further considered here), the installation of hi-fi and computers will often have to be done by the supplier. As always, this should be realistically priced. If the customer suggests that he/she can buy the boxes elsewhere for a lower price, point out by all means that you will supply an installation service at a price that reflects a fair return, and that such installation is an integral part of the price that you quote for equipment.

Outside installation follows the same pattern as was mentioned earlier for outside repairs, but with no option for taking the equipment to the workshop. The customer needs to be consulted about positioning of equipment – the loudspeaker may, after all, clash with the colour of the curtains – but the installer should be able to point out tactfully that a full stereo effect will not be experienced if the loudspeakers are placed close to each other, facing in different directions or even (as I have seen) in different rooms. Similarly, it is up to the installer to show that the monitor of a computer should be placed where the user of the keyboard can see it without needing to lean forward or sideways, that the mouse should come conveniently to hand, and that the flow of air through the main casing is unobstructed.

Installation for white goods depends critically on the presence of electrical sockets, water supplies and drains, and customers should understand that alterations to the house wiring and/or plumbing are not undertaken by the service engineer who is installing equipment.

Though the interconnection of equipment is often so standardized that you can work without manuals, you should be aware of any peculiarities of equipment, limitations to cable lengths (particularly for parallel printers for computers), etc. A useful hint for computers that are used with a large amount of powered peripherals (such as scanner, modem, printer, etc.) is to connect the power take-off socket on the main computer case to a set of distribution sockets, which can be used to supply to peripherals. This allows the user to switch everything on and off together using only the main switch on the computer. A similar scheme can be used for the more elaborate type of hi-fi setup.

Never leave an installation without testing it adequately – do not assume that because it sounds good playing a cassette that the CD player will be as good. In some cases, you may need, for example, to alter sensitivity levels at inputs. In addition, never leave an installation without making sure that the customer knows how to operate the equipment and what peculiarities of installation are present. Few users, for example, know how to deal with the connections between a modern VCR and TV receiver, and many assume that video replay still needs the use of a spare channel rather than the AV option.

EMC – CE marking

Electromagnetic compatibility is defined as the ability of a device, equipment or system to function satisfactorily in its electromagnetic environment without introducing intolerable electromagnetic interference to any other system, while at the same time, its own performance will not be impaired by interference from other sources. Since interference is a destroyer of information, it is often a limiting factor in a communications system. For convenience, interference can be divided into two categories, natural and man-made. The former often results in electrostatic discharge (ESD), while the latter usually results in power supply line surges. Natural sources of interference include ionospheric storms and lightning. Man-made interference commonly results from high current switching operations and radio frequency generating equipment, including industrial, scientific and medical (ISM) apparatus. With the expansion in the use of portable digital communications equipment such as computers and cordless telephones, and the introduction of information technology (IT) systems into offices and the domestic environment, the sources of interference are growing rapidly.

White goods seldom suffer from interference, but can be a cause of interference to other electronic equipment. With the increasing use of microprocessor controls in white goods, however, interference can become as serious a problem as it is for brown goods.

A direct lightning strike is not necessary to cause havoc in a communications system. Even a cloud-to-cloud discharge can easily set up an electric field of the order of 50 volts per metre, at a distance of 1 kilometre. Such a situation could induce a voltage as high as 50 kV on a kilometre of exposed power line or telephone cable. Fortunately, such induced surges are usually limited to very short periods of time.

The action of a person walking across a carpet or even sitting in a chair can generate very high electrostatic charges, depending upon clothing and the level of humidity. This *triboelectricity* due to rubbing dissimilar materials together can often be the source of computing disasters. A typical body capacitance can range between 100 and 250 pF and acquire a charge of 0.1 to 5 µC. With a human body discharge resistance of around 150 to 1500 ohms and taking midrange values of 150 pF and 3 µC, this can produce a voltage of around 13.3 kV ($Q = CV$) with an energy of about 13.3 mJ ($E = 0.5CV^2$). (Because of the $Q = CV$ relationship any movement such as lifting the feet off the floor which lowers the body capacitance will automatically increase the voltage level.) If this is discharged through a typical body resistance in from 0.5 to 20 ns, the peak current can be in the order of 1 to 50 A.

Even if exposures to such discharges do not prove to be immediately destructive, they can introduce a latent failure mechanism for microelectronic circuits because these effects are cumulative.

A triboelectric series lists the materials ranging from air, hands, asbestos, at the most positive, through the metals, to silicon and Teflon as the most negative. The further apart in this table, then the higher will be the electric charge between two materials when rubbed together.

While the British Standards Institution (BSI) provides some standards for EMC/EMI that include domestic, industrial and medical applications, the commonly quoted European organizations are CISPR (Comité International Special des Perturbations Radioelectriques) which is a committee of the IEC (International Electrotechnical Commission) and VDE (Verband Deutsche Elektrotechniker).

From 1st January 1996, all electrical all electrical or electronic equipment must comply with the European Directive on EMC, 89/336/EEC. A product can only be sold in Europe if it meets the requirements of the directive. The few exceptions include very low power or battery-operated equipment and toys. If a complaint arises and the product is found to be non-compliant, it will be banned from the European Union (EU) and European Free Trade Area (EFTA) countries. All compliant equipment must carry the CE mark. The relevant standards supporting the directive are drafted in the CENELEC Generic Standards EN50081-1 and EN50082-l for emission and immunity respectively.

From the 1st January 1997, the Low Voltage Directive (LVD), 73/23/EEC becomes effective for equipment connected to a mains supply. In this case, low voltage refers to values below 1 kV a.c. and 1.5 kV d.c. The standard differs from the EMC directive in a number of ways:

- It dates from 1973.

- It is much more complex.

- It effectively provides an extension of the CE marking scheme for EMC.

- Third party testing routes for compliance are through BABT for telecommunications and BEAB for consumer equipment.

Recent changes to British Electrotechnical Approvals Board (BEAB) standards

The Consumer Electronics Standard IEC60065:1998 which is now in force sets the relevant test standards for CE marked equipment. Whilst this regulation applies particularly to manufacturing, there are implications for the service technician. The most important additions apply to flammability and affects such as equipment covers, PCBs, mains switches and high voltage components (greater than 4 kV) which now have to meet new standards of flame retardancy.

It is now necessary to consider potential ignition sources and barriers to resist the spread of fire. Material flammability is covered by the Flammability Standard IEC60707 which is expected to become fully operative during August 2002. This specifies the distances between potential ignition sources and a combustible component. This distance ranges from 13 mm to more than 50 mm where a voltage greater than 4 kV exists, such as in a television receiver. Where these distances cannot be achieved the region must be divided by a non-combustible barrier. In regions such as around the line output stage of a TV receiver where it is impossible to meet these requirements, it is possible to fence the region within an internal fire enclosure.

Basically these standards impose design problems for the manufacturers but it is incumbent upon the service technician to maintain them. By way of a warning, it might usefully be pointed out that the On/Off switches of a number of TV receivers that were operated in the *Standby* mode for long periods, have been known to overheat and become the cause of several serious domestic fires.

Planned or managed maintenance

System servicing and maintenance normally operates within the bounds of two extremes. 'If it ain't broke, don't fix it' and 'If anything can go wrong it will and always at the most inconvenient time'. If a TV receiver or CD player fails, it is nothing more than an inconvenience, but when the monitor or CD-ROM of a computer control system goes down, the effect becomes catastrophic.

Unexpected failures within a production or control environment can create loss of production or service which leads to loss of customer confidence. This is most important in 'just in time' (JIT) applications where there is virtually no buffer stock of components to allow production to continue. Out of hours repairs by highly qualified service technicians/engineers, not only adds to system/production costs, but also increases the health and safety hazards. Responding to unexpected failures is often described as 'fire-fighting'.

The basis for a planned maintenance scheme is built on the system manufacturers' recommendations and requires a significant constant retraining element to ensure that all service personnel can respond quickly and accurately to any indication that a failure is imminent. An early response and a temporary fix can often allow proper maintenance to be initiated at a more convenient slack period.

Service documentation for a system needs to be constantly updated in line with operation experience. Data logging is therefore essential and many computer software packages are available to support this effort. Over a period of time it becomes possible to predict areas of the system that need more regular service attention so that historical records become very important. System down time costs money and loses orders.

For the larger systems that operate on a 24 hours a day basis the annual holiday period with its complete shut down provides something of a respite for service personnel. Maintenance is then not just a case of fire fighting. Smaller systems may have to be serviced over night. Maintenance of a major nature thus has to be fitted in with predicted non-operation times.

When an unplanned halt occurs, it might provide a useful period in which to carry out service on other parts of the system, particularly those areas nearest due for planned maintenance. Thus making use of such down time to minimize the duration of planned breaks at a later date.

Planned maintenance normally operates on a time basis and many machines actually carry a run time meter to help. Some areas need daily attention for say lubrication and running adjustments, whilst other areas need daily, weekly or even annual attention. With such a variability within a system maintenance scheme, it is important that a readily available and up-to-date database is maintained. The recording of failures should lead to the pin pointing of stock faults which might then be minimized by more regular servicing.

The most important features of a planned maintenance scheme can now be restated and include;

- obtaining a thorough understanding of the system operation,
- continuous re-training for personnel,
- accurate maintenance of historical records,
- management plan modified in the light of experience,
- ensure that maintenance routines are written in clear and understandable manner.

Answers to questions

1.1   Electrical, fire, asphyxia, mechanical.
1.2   Low voltage, low current, dry hands, dry floor and rubber-soled shoes, work single-handed.
1.3   Paper, wood, cloth, but never on electrical equipment.

2   D.c. technology

Electrical units

Electric current consists of the flow of small particles called electrons in a circuit. Its rate of flow is measured in units called *amperes*, abbreviated either to 'amps' or 'A'. One ampere is the amount of electric charge, in units called *coulombs*, which passes a given point in a circuit per second. The coulomb has a value of about 6.289×10^{18} electrons. The measurement of current is done, not by actually counting these millions of millions of millions of electric charges, but by measuring the amount of force that is exerted between a magnet and the wire carrying the current which is being measured.

The ampere is a fairly large unit, and for most electronics purposes the smaller units milliamp (one thousandth of an ampere) and microamp (one millionth of an ampere) are more generally used. The abbreviations for these qualities are mA and μA respectively.

In the same way as a pressure is needed to cause a flow of water through a pipe, so an electrical pressure known as electromotive force (also called a *voltage*) is needed to push a current through a resistance. Electromotive force (emf) is measured in units of *volts*, symbol V. A voltage is always present when a current is flowing through a resistance, and the three quantities of volts, amps and ohms are related.

Circuits and current

An electric *circuit* is a closed path made from conducting material. When the path is not closed, it is an *open circuit*, and no current can flow. In a circuit that contains a battery and a lamp, for example, the lamp will light when the circuit is closed, and we take the direction of the flow of current as from the positive (+) pole of the battery to the negative (−).

We can show an arrangement such as this in two ways. One is to draw the battery and the bulb as they would appear to the eye. The other is to draw the shape of the circuit, representing items such as the battery and the lamp as symbols, the components of the circuits, and the conductor as a line.

We draw these *circuit diagrams* to show the path that the current takes, because this is more important than the appearance of the components. To avoid confusion, there are some rules (conventions) about drawing these circuits.

- A line represents a conductor
- Where lines cross, the conductors are *NOT* joined
- Where two lines met in a T junction with or without a dot, conductors *are* connected.

Figure 2.1 shows some symbols that are used for common components. Most of these use two connections only, but a few use three or more. These symbols are UK and European standards, but circuit diagrams from USA and Japan may use the alternate symbols for resistors and capacitors.

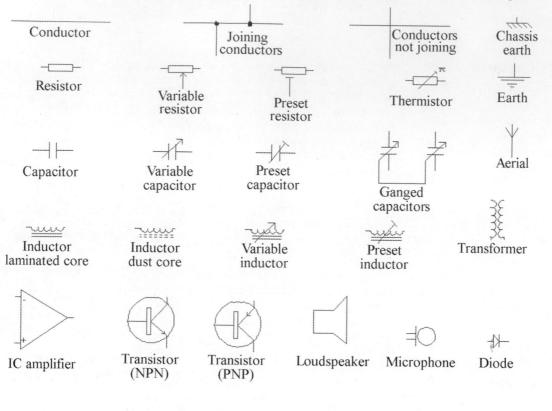

Figure 2.1   Some symbols used in circuit diagrams

You will still find circuits that are old or of foreign origin, that show crossing lines meaning joined conductors. On these older drawings, a loop is used to indicate that conductors are not joined.

Circuit diagrams are important because they are one of the main pieces of information about a circuit, whether it is a circuit for the wiring of a house or the circuit for a TV receiver. For servicing purposes you must be able to read a circuit diagram and work out the path of currents.

A.c., d.c. and signals

Electrical current can be *d.c.* or *a.c.*, and mixtures of both. D.c. means *direct current*, and it is a steady flow of current. This is the type of current that you get in a battery circuit, for example, and d.c. is also used to operate electronic equipment. A.c. means *alternating current*, meaning that the current is not steady; its value rises, falls, reverses, and rises and falls in the reverse direction. A.c. current is a form of wave, and when the voltage or current of mains a.c. is plotted on a graph it provides the shape of a sine wave. A.c. is generated by rotating machines such as alternators, and by electronic circuits called oscillators. We shall look more closely at a.c. in Unit 2.

Effects of current

Electric current causes three main effects which have been known for several hundred years.

- **Heating effect:** when a current flows through a conductor, heat is generated so that the temperature of the conductor rises.
- **Magnetic effect:** when a current flows through a conductor it causes the conductor to become a magnet.
- **Chemical effect:** when a current flows through a chemical solution it can cause chemical separation.

All of these effects can be either useful or undesirable. We use the heating effect in electric fires and cookers, but we try to minimize the loss of energy from transmission cables by using high voltages with low current for transmission. The heating effect is the same whether the current is d.c. or a.c.

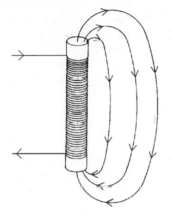

The magnetic effect of d.c. can be increased by winding the wire into a coil (see left), a *solenoid*. The effect of using alternating current is to produce magnetism which changes direction every time the current through the wire changes direction. The magnetic effect is used in electric motors, relays and solenoids (meaning magnets that can be switched on and off). Less desirable effects include the unwanted interference that comes from the magnetic fields created around wires.

One notable wanted chemical effect is that of chemical energy is converted to electrical energy in a cell or battery. Some other chemical effects are, however, undesirable. A current that is passed through a solution of a salty material dissolved in water will cause a chemical change in the solution which can release corrosive substances. This is the effect that causes *electrolytic corrosion*, particularly on electrical equipment that is used in ships.

All of these effects have been used at one time or another to measure electric current. We now use the magnetic effect in the older type of instruments, but the modern digital meters work on quite different principles.

Question 2.1

Why is it better to transmit electric power using a high voltage and low current rather than a low voltage and high currrent?

Denary numbers

A *denary number* is one which is either a multiple of ten, like 2, 50, or 350, or a fraction that has 10 or a multiple of ten (*power* of ten) below the bar, like $^1/_{10}$, $^3/_{40}$ or $^7/_{120}$. The number 100 is a denary number, equal to 10 tens, and we count in multiples (*powers*) of ten, using 1000 ($10 \times 10 \times 10$),

10 000, 100 000 and so on. The number 89 is a denary number, equal to eight tens plus nine units. The number 0.2 is also a denary number equal to $^2/_{10}$, a decimal *fraction*.

Denary numbers can be added, subtracted, multiplied and divided digit by digit, starting with the *least significant* figures (the units of a whole number, or the figure furthest to the right of the decimal point of a fraction), and then working left towards the *most significant* figures.

Decimals are denary numbers that are fractions of ten, so that the number we write as 0.2 means $^2/_{10}$, and the number we write as 3.414 is $3 + {}^{414}/_{1000}$. The advantage of using decimals is that we can add, subtract, multiply and divide with them using the same methods as we use of whole numbers. Even the simplest calculators can work with decimal numbers.

Powers of ten: significant figures

The numbers 0.047, 47 and 47 000 are all denary numbers. Each of them consists of the two figures 4 and 7, along with a power of ten which is shown by zeros put in either before or after the decimal point (or where a decimal point would be). The number 0.047 is the fraction 47/1 000, and 47 000 is 47 × 1 000. The figures 4 and 7 are called the *significant figures* of all these numbers, because the zeros before or after them simply indicate a power of ten. *Zero can be a significant figure* if it lies between two other significant figures as, for example, in the numbers 407 and 0.407. The zeros in a number are not significant if they follow the significant figures, as in 370 000, or if they lie between the decimal point and the significant figures, as in 0.000 23.

For many purposes in electronics, only two significant figures are needed to express the value of a quantity, so that numbers can often be expressed as two figures multiplied by a power of ten, either larger or smaller than unity. The power of ten is the number of zeros, so that 1000 is ten to the power 3, written as 10^3 and 1/1000 is ten to the power −3, written as 10^{-3}.

Table 2.1   Powers of ten in index form

Number	Power	Written as:
1/1 000 000 or 0.000001	−6	10^{-6}
1/100 000 or 0.00001	−5	10^{-5}
1/10 000 or 0.0001	−4	10^{-4}
1/1 000 or 0.001	−3	10^{-3}
1/100 or 0.01	−2	10^{-2}
1/10 or 0.1	−1	10^{-1}
1	0	10^{0}
10	1	10^{1}
100	2	10^{2}
1 000	3	10^{3}
10 000	4	10^{4}
100 000	5	10^{5}
1 000 000	6	10^{6}

Powers of ten are always written in this *index* form, as shown in table 2.1. A positive index means that the number is greater than one (*unity*), and a negative index means that the number is less than unity. For instance, the number $1.2 \times 10^3 = 1\,200$, the number $47 \times 10^{-2} = 4.7 \times 10^3 = 0.047$, and so on.

There is an easier way of using powers of ten when we are writing the value of any electrical units such as volts (V), amperes (I) or ohms (R). This is done by using a system of standard *prefixes*, meaning letters that come ahead of the unit name, like *milli*volts or *kilo*hms. Table 2.2 lists some of the more commonly used of these prefixes. The multiplier figures have been omitted when the number of zeros exceeds eleven. The more commonly used prefixes are those that lie between pico and giga.

Table 2.2   Some standard prefixes

Multiplier	Power	Prefix	Abbreviation
	10^{-24}	yocto	y
	10^{-21}	zepto	z
	10^{-18}	atto	a
	10^{-15}	femto	f
0.000 000 000 001	10^{-12}	pico-	p
0.000 000 001	10^{-9}	nano-	n
0.000 001	10^{-6}	micro-	μ
0.001	10^{-3}	milli-	m
1 000	10^3	kilo-	K or k
1 000 000	10^6	mega	M
1 000 000 000	10^9	giga-	G
1 000 000 000 000	10^{12}	tera-	T
	10^{15}	peta	P
	10^{18}	exa	E
	10^{21}	zetta	Z
	10^{24}	yotta	Y

Two simple examples will help to show how the system works. A current flow of 0.015 amperes can be more simply written as 15 mA (milliamps), which is 15×10^{-3} A. A resistance of 56 000 ohms, which is equal to 56×10^3 ohms, is written as 56K (K for kilohms). The ohm sign Ω is often left out.

The *British Standard* (BS) system of marking values of resistance (BS1852/1977) uses this method, but with a few changes. The main difference is that the ohm sign Ω and the decimal point are *never* used. This avoids making mistakes caused by an unclear decimal point, or by a spot mark mistaken for a decimal point. This is particularly important for circuit diagrams that are likely to be used in workshop conditions.

In this BS system, all values in ohms are indicated by the letter R, all values in kilohms by the capital letter K, and all values in megohms by M. These letters are then placed where the decimal point would normally be found, and the latter is thereby eliminated. Thus R47 = 0.47 ohms; 5K6 = 5.6 kilohms; 2M2 = 2.2 megohms. The BS system is illustrated throughout this book. In this system there is no space between the number and the letter.

Question 2.2

The BS value system is used also for capacitance values and for some voltage values such as the stabilized value of a Zener diode.

Write in amps the following quantities: 5 mA, 47 μA, 221 nA. You can use normal decimal notation or power-of-ten notation.

Calculators, simple and scientific

At one time we used tables of values to help in solving complicated calculations, or calculations that used numbers containing many significant figures. Nowadays we use electronic calculators in place of tables, but a calculator is useful only if you know how to use it correctly. Calculators can be simple types that can carry out addition, subtraction, multiplication and division only, and these can be useful for most of your calculations.

To solve some of the other types of calculations you will meet in the course of electronics servicing, a *scientific calculator* is more useful. A good scientific calculator, such as the Casio, need not be expensive and it will be able to cope with any of the calculations which will need to be made throughout this course. You should learn from the manual for your calculator how to carry out calculations involving square roots and powers, sines and cosines, and the use of brackets.

Ratios

Many of the quantities used in electronics measurements are ratios, such as the ratio of the current (I_c) flowing in the collector circuit of a transistor to the current (I_b) flowing in its base circuit.

A ratio consists of one number divided by another, and can be expressed in several different ways:

$1/12 = 0.083 = 8.3\%$

- As a common fraction, such as 2/25.

- As a decimal fraction, such as 0.47.

- As a percentage, such as 12% (which is another way of writing the fraction 12/100).

To convert a decimal fraction into a common fraction, first write the figures of the decimal, but not the point. For example, write 0.47 as 47. Now draw a fraction bar under this number (called the numerator) and under it write a power of ten with as many zeros as there are figures above. In this example, you would use 100, with two zeros because there are two figures in 47. This makes the fraction $^{47}/_{100}$.

To convert a common fraction into a decimal, do the division using a calculator. For example, the fraction 2/27 uses the 2, division and 27 keys and comes out as 0.074 074, which you would round to 0.074.

To convert a decimal ratio into a percentage, shift the decimal point two places to the right, so that 0.47 becomes 47%. If there are empty places, fill them with zeros, so that 0.4 becomes 40%.

To convert a percentage to a decimal ratio, imagine a decimal point where the % sign was, and then shift this point two places to the left, so that 12% becomes 0.12. Once again, empty places are filled with zeros, so that 8% becomes 0.08.

Decibel scales

The human ear does not respond to increases or decreases of sound power in a linear manner. For example, if we create a series of doublings of power output levels say 1, 2, 4, 8, 16 watts, etc., the ear would interpret this as a linear scaling which it relates to the power or logarithm of the number (i.e. $2^0, 2^1, 2^2, 2^3, 2^4$, etc.). It is therefore more convenient to measure changes of intensity or power using a logarithmic scale using the relationship as the logarithm of the power ratio.

This ratio has been defined as the bel (B) which has a relatively large value so that it is common to use the decibel (dB) which is one tenth of a bel. The formula now becomes, $10 \times \log$ (power ratio) or more commonly $10 \log(P_1/P_2)$. This unit is also more convenient because the smallest change in level that the ear can detect is about 1dB.

If the power levels in question are developed across or in the same value of resistance, then, because power is proportional to V^2 or I^2, it becomes possible to relate V or I on a similar scale as $dB = 20 \log(V_1/V_2)$. Quantities such as system signal to noise ratios may also be conveniently expressed in dB.

A doubling of power can be expressed as $10 \log 2 = 10 \times 0.30 = 3.01$ dB. Because fractional dB are not normally audible, this is usually rounded down to 3 dB. However, in complex systems where the effects of noise etc are additive, it may be useful to work to values as low as 0.1 dB. Losses in a system are usually represented by negative values of decibels.

The doubling of voltage or current levels yields the following, $20 \log 2 = 6.02$ dB (or when rounded down) = 6 dB. Since the doubling of V or I in the same value of resistance, is equivalent to quadrupling the power relationship, this is again $2 \times 3 = 6$ dB.

The system of decibels can be calculated using the inverse process as follows, 1 dB $= 10 \log$(power ratio), so that the power ratio = the antilog $(1/10) = 1.259$. (For antilog read 10 to the power 0.1.) Therefore an increase or decrease of 1 dB is equivalent to a change in power level of about 26%. In a similar way, the ln and e^x are inverse functions, but to the base e rather than 10. (e is approximately equal to 2.7183.) On your calculator, the log is obtained by keying in the ratio value (such as 2, 5, etc.) and then pressing the log key.

A decibel scale is normally used when the gain of an amplifier is plotted against frequency, see later. The decibel corresponds to the smallest change that we can detect when we are listening to the amplifier.

Averages and tolerances

The *average value* of a set of numbers is found by adding up all the numbers in the set and then dividing by the number of items in the set. Suppose that a set of resistors has the following values: one 7R, two 8R, three 9R, four 10R, four 11R, three 12R, and two 13R. This is a set of 19 values, and the average value of the set is found as follows:

$$\frac{7+8+8+9+9+9+10+10+10+10+11+11+11+11+12+12+12+13+13}{19} = \frac{196}{19}$$

This divides out to 10.32, so that the average value of the set is 10.32 ohms or 10R32.

An average value like this is often not 'real', in the sense that there is no actual resistor in the set that has the average value of 10.32R. It is like saying that the average family size in the UK today is 2.2 children. This might be a perfectly truthful average value statement but you will seldom meet a family containing two children and 0.2 of a third one.

In a manufacturing process, the average value is a target value, one that the manufacturer aims to achieve. This might be a value of length, of volume, of electrical resistance or of anything else. When the value of a component turns out to be more or less than the target value, the component is said to have a *tolerance*. Thus in a box containing 10R resistors, a resistor might have a measured value of 12R. The resistor has a tolerance of 2R. Tolerances are in practice usually written as percentages of the target value. In this example, 2R/10R expressed as a percentage becomes $2/10 \times 100\% = 20\%$, so that the tolerance of the odd resistor is within 20% of the target value.

A tolerance of 20% means that some samples in a batch may be as much as 20% high (+) and others 20% low (–) compared to the target or average value of the batch as a whole. If you had a box of 120 ohm resistors with 20% tolerance you would expect to find resistors of 96 ohms and of 144 ohms. This is because 20% of 120 is 24. Subtracting, 120 – 24 = 96, and adding gives 120 + 24 = 144. You would not find any resistor of 120 ohms (or even close to 120 ohms) because the manufacturer has sorted the resistors by value and put all of the resistors that are more than 10% out of tolerance into the 20% set.

Question 2.3

What is the average of 27, 33, and 47?

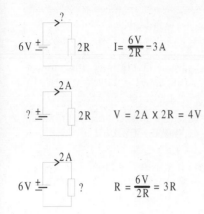

$$I = \frac{6V}{2R} = 3A$$

$$V = 2A \times 2R = 4V$$

$$R = \frac{6V}{2R} = 3R$$

The electrical units of volts, amps and ohms are related, and the relationship is commonly known (not quite correctly) as *Ohm's law*, which as an equation is written as $V = R \times I$. In words, it means that the voltage measured across a given resistor (in volts) is equal to the value of the resistance (in ohms) multiplied by the amount of current flowing (in amperes).

The Ohm's law equation can be rearranged, as illustrated, as either $R = V/I$ (resistance equals volts divided by current) or $I = V/R$ (current equals volts divided by resistance).

These equations are the most fundamentally important ones you will meet in all your work on electricity and electronics. In electrical circuits the units in which the law has been quoted volts, amperes, ohms should normally always be used; but in electronic circuits it is in practice much easier to measure resistance in K and current in mA. Ohm's law can be used in any of its forms when both R and I are expressed in these latter units, but the unit of voltage in these other expressions remains always the volt.

There are, therefore, two different combinations of units with which you can use Ohm's law as it stands:

either: VOLTS AMPERES OHMS

or VOLTS MILLIAMPERES KILOHMS

Never mix the two sets of units. Do not use milliamperes with ohms, or amperes with kilohms. If in doubt, convert your quantities to volts, amps and ohms before using Ohm's law.

The diagrams illustrate how Ohm's law is used. Here are four more difficult examples:

Example: What is the resistance of a resistor when a current of 0.1 A causes a voltage of 2.5 V to be measured across the resistor?

Solution: Express Ohm's law in the form in which the unknown quantity R is isolated: R = V/I. Substitute the data in units of volts and amperes.

R = 2.5/0 1 = 25R.

Example: What value of resistance is present when a current of 1.4 mA causes a voltage drop of 7.5 V?

Solution: The current is measured in milliamps, so the answer will appear in kilohms. R = VI = 7.5/1.4 = 5.36 kilohms, or about 5K4.

Example: What current flows when a 6K8 resistor has a voltage of 1.2 V across its terminals?

Solution: The data are already in workable units, so substitute in I = V/R Then I = 1.2/6.8 A = 0.176 A, or 176 mA.

Example: What current flows when a 4K7 resistor has a voltage of 9 V across its terminals?

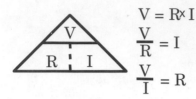

$$V = R \times I$$

$$\frac{V}{R} = I$$

$$\frac{V}{I} = R$$

Solution: With the value of the resistor quoted in kilohms, the answer will appear in milliamps. So substitute in $I = V/R$, and I = 9/4.7 = 0.001 915 A = 1.915 mA.

The importance of Ohm's law lies in the fact that if only two of the three quantities current, voltage and resistance are known, the third of them can always be calculated by using the formula.

The important thing is to remember which way up Ohm's law reads. Draw the triangle illustrated here. Put V at its **V**ertex, and I and R down below and you will never forget it. The formulae follow from this arrangement automatically using a 'cover-up' procedure. Place a finger over I and V/R is left, thus I = V/R. The other ratios can be found in a similar way.

Work, power and energy

These three related quantities are often confused. Mechanical work is done whenever a force F is overcome or moved through a distance d. The force is measured in newtons N and 1N is the force necessary to produce an acceleration of 1 metre per second per second (1 m/s^2) to a mass of 1 kilogram. Work is therefore the product of F × d newton metres, and this unit is called the joule. Work is also directly related to the torque or turning moment applied to a shaft.

Power is the rate at which work is done and electrically is measured in watts or joules (the unit of work) per second. Work also generates heat and this is related by the equation, 4.2 joules per calorie, where 1 calorie is the heat required to raise the temperature of 1 gram of water (1cc) through 1°C. Electric motors were often specified by their work loading in horse power or brake horse power (HP or BHP) on the rating plate, where 1 HP is equivalent to 746 watts. Therefore a ½ HP AC motor would draw just over 1.5 amps from the nominal 240 volts supply mains.

Energy is the capacity to do work and is the product of power and time. Thus 1 joule is equal to 1 watt second. It can then easily be shown that 1 kWh = 3 600 kJoules (kJ) or 3.6 MJoules (MJ). The term *shaft horsepower* is sometimes encountered and this applies to a system whereby the actual driving mechanism is coupled to the motor via a gearbox or similar gearing. A step down gear ratio unit is often described as a torque multiplier, thus a 2:1 step down ratio driven from a motor of 100 HP would produce an output of 200 shaft HP.

Question 2.4

The power of a car engine is often quoted in kW as well as in BHP. What BHP corresponds to a rating of 85 kW?

Electrical power

Power = $V \times I$ watts

= V^2/R watts

= I^2R watts

When a current flows through a resistor, electrical energy is converted into heat energy, and this heat is passed on to the air around the resistor, and dissipated, spread around. The rate at which heat is dissipated is called *power*, and is measured in units of watts.

The amount of power dissipated can be calculated from any two of the quantities V (in volts), I (in amps) and R (in ohms), as follows:

Using V and I	Power = $V \times I$ watts
Using V and R	Power = V^2/R watts
Using I and R	Power = I^2R watts

Most electronic circuits use small currents measured in mA, and large values of resistance measured in K, and we seldom know both volts and current. The power dissipated by a resistor is therefore often more conveniently measured in milliwatts using volts and K or using milliamps and K. Expressing the units V in volts, I in milliamps and W in milliwatts, the equations to remember become:

The milliwatts dissipated = V^2/R (volts and K)

= I^2R (milliamps and K)

Example: How much power is dissipated when:

(a) 6 V passes a current of 1.4 A,

(b) 8 V is placed across 4 ohms,

(c) 0.1 A flows through 15R?

Solutions:

(a) Using $V \times I$ Power = $6 \times 1.4 = 8.4$ W

(b) Using V^2/R Power = $8^2/4 = 64/4 = 16$ W

(c) Using I^2R Power = $0.1^2 \times 15 = .01 \times 15 = 0.15$ W

Example: How much power is dissipated when:

(a) 9 V passes a current of 50 mA?

(b) 20 V is across a 6K8 resistor?

(c) 8 mA flows through a 1K5 resistor?

Solution:

(a) Using $V \times I$ Power = $9 \times 50 = 450$ mW

(b) Using V^2/R Power = $20^2/6.8 = 400/6.8 = 58.8$ mW

(c) Using I^2R Power = $8^2 \times 1.5 = 64 \times 1.5 = 96$ mW

The amount of energy which is dissipated in the manner described is measured in joules. The watt is a rate of dissipation equal to the energy loss of one joule per second, so that joules = watts × seconds or watts = joules/seconds. The energy is found by multiplying the value of power dissipation by the amount of time during which the dissipation continues. The resulting equations are: Energy dissipated = $V \times I \times t$ joules or V^2t/R joules or I^2Rt joules where t is the time during which power dissipation continues, measured in seconds. In electronics you seldom need to make use of joules except in heating problems, or in calculating the stored energy of a capacitor.

Electrical components and appliances are rated according to the power which they dissipate or convert. A 3 W resistor, for example, will dissipate 3 joules of energy per second; a 3 kW motor will convert 3000 joules of energy per second into motion (if it is 100 per cent efficient).

As a general rule, the greater the power dissipation required, the larger the component needs to be.

Cells and batteries

Cells convert chemical energy into d.c. electrical energy without any intermediate stage of conversion to heat. Only a few chemical reactions can at present be harnessed in this way, though work on fuel cells has enabled electricity to be generated directly without any fuel having to be burned to provide heat. Cells and batteries, however, though important as a source of electrical energy for electronic devices, represent only a tiny (and expensive) fraction of the total electrical energy which is generated.

A *cell* converts chemical energy directly into electrical energy. A collection of cells is called a *battery*. Cells may be connected in series to increase the voltage available or in parallel to increase the current capacity, but parallel connection is usually undesirable because it can lead to the rapid discharge of all cells if one becomes faulty and the others pass current into the faulty cell.

Cells may be either primary or secondary cells. A *primary cell* is one that is ready to operate as soon as the chemicals composing it are put together. Once the chemical reaction is finished, the cell is exhausted and can only be thrown away. A *secondary cell* generally needs to be charged before it can be used. Its chemical reaction takes place in one direction during charging, and in the other direction during discharge (use) of the cell.

Most primary cells are of the zinc/carbon (Leclanché) type, of which a cross-section is shown in Figure 2.2. The zinc case is sometimes steel coated to give extra protection. The ammonium chloride paste is an acidic material which gradually dissolves the zinc. This chemical action provides the energy from which the electrical voltage is obtained, with the zinc the negative pole.

The purpose of the manganese dioxide depolarizer mixture that surrounds the carbon rod is to absorb hydrogen gas, a by-product of the chemical reaction. The hydrogen would otherwise gather on the carbon, insulating it so that no current could flow. The zinc/carbon cell is suitable for most purposes for which batteries are used, having a reasonable shelf life and yielding a fairly steady voltage throughout a good working life.

Other types of cell such as alkaline manganese, mercury or silver oxide and lithium types are used in more specialized applications that need high working currents, very steady voltage or very long life at low current drains. However, mercury based cells are not considered environmentally friendly when discarded unless they can be returned to the manufacturer. The use of a depolarizer is needed only if the chemical action of the cell has generated hydrogen, and some cell types do not.

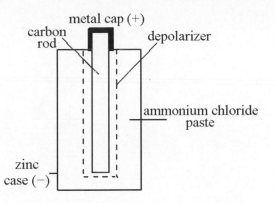

Figure 2.2   A typical (Leclanché) dry cell construction

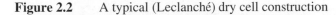

Activity 2.1

Connect the circuit of Figure 2.3 using a 9 V transistor radio battery. Draw up a table on to which readings of output voltage V and current I can be entered.

With the switch Sw1 open, note the voltmeter reading (using the 10 V scale). Mark the current column 'zero' for this voltage reading. Then close Sw1 and adjust the variable resistor until the current flow recorded on the current meter is 50 mA. Note the voltage reading V at this level of current flow, and record both readings on the table. Open switch Sw1 again as soon as the readings have been taken.

Go on to make a series of readings at higher currents (75 mA, 100 mA, etc.) until voltage readings of less than 5 V are being recorded. Take care that for every reading Sw1 remains closed for only as long as is needed to make the reading. Plot the readings you have obtained on a graph of output voltage against current. It should look like the example shown in the diagram, left.

Now pick from the table a pair of voltage readings V_1, and V_2, with V_2 greater than V_1, together with their corresponding current readings, I_1 and I_2, expressed in amperes. Work out the value of the expression shown, left, and you will get the internal resistance of the battery in units of ohms.

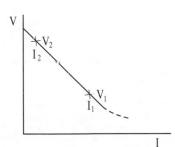

$$\frac{V_2 - V_1}{I_2 - I_1}$$

- Note that false readings can be obtained if a cell passes a large current for more than a fraction of a second. Try to make your readings quickly when you are using currents approaching the maximum, and switch off the current as soon as you have taken a reading.

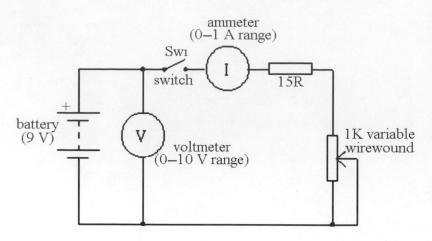

Figure 2.3   Diagram for Activity 2.1

Most primary cells have an open-circuit voltage (or EMF) of around 1.4 –1.5 V. The important exception is the lithium cell, see later, which provides around 3.0 V. A lithium cell must *never* be opened, because lithium will burst into flames on exposure to air or water. Lithium cells should not be re-charged, nor put into a fire.

Towards the end of the useful life of a cell or battery, the value of its *internal resistance* rises. This causes the output voltage at the terminals of the cell or battery to drop below its normal value when current flows through the cell or battery which is then said to have poor *regulation*. A voltage check with this cell or battery removed from the equipment will show a normal voltage rating, but the cell/battery should nevertheless be replaced.

> The only useful check on the state of a cell or battery is a comparison of voltage reading *on load* (with normal current flowing) with the known on-load voltage of a fresh cell. Simply reading the voltage of a cell that is not connected to a load is pointless.

At one time, the term secondary cell meant either the type of lead-acid cell which is familiar as the battery in your car, or the nickel-iron alkaline (NiFe) cell used in such applications as the powering of electric milk floats. In present-day electronics, both types have to some extent been superseded by the nickel-cadmium (NiCd or Nicad), nickel-metal-hydride (Ni-MH), and lithium-ion secondary (meaning rechargeable) cells. There is, however, a large difference in the EMF of secondary cells. The old lead-acid type has an EMF of 2.0 V (2.2 V when fully charged), but the nickel cadmium and Ni-MH types have a much lower EMF of only 1.2 V, and the lithium ion cell can provide around 3.6 V.

As a matter of interest, both types of alkaline cells (because they use alkaline rather than acid electrolyte) were invented in the same year, 1900. The nickel iron cell was invented by Edison in the USA, and the nickel cadmium cell by Jungner in Sweden. The advantages of the nickel-cadmium cell have been exploited only recently, however. The active material cadmium (a metal like to zinc), in powdered form, is pressed or sintered into perforated steel plates, which then form the negative pole of the cell. The positive pole is a steel mesh coated with solid nickel hydroxide. The electrolyte is potassium hydroxide (caustic potash), usually in jelly form (see Figure 2.4).

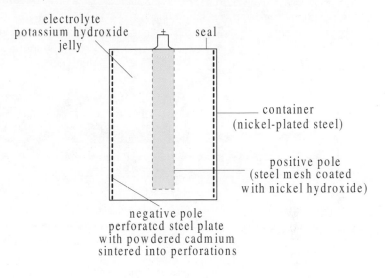

Figure 2.4   The nickel-cadmium cell

Nickel-cadmium (nicad) cells are sealed so that no liquids can be spilled from them, and they have a fairly long working life provided they are correctly used. They can deliver large currents, so they can be used for equipment that demands higher power than could be supplied by primary cells. They have much longer life than other cells in applications for which they are rapidly discharged at intervals. Long periods of inactivity can cause the cells to fail, though it is often possible to restore their action by successive cycles of discharging and charging. One major problem is the memory effect, a reduction in capacity caused by recharging before the cell is fully discharged. Because of this, it is common practice to use a discharging cycle before charging a nickel-cadmium cell.

More recently, the sealed nickel-metal hydride (Ni-MH) battery has been introduced. This type has up to 40% higher capacity than its nickel-cadmium counterpart of the same size, and also offers benefits of faster charge and discharge rates, and longer life. The Ni-MH cell contains no cadmium and is therefore more environmentally acceptable. The operating voltage is about the same as that of the Ni-Cd cell. The memory effect can be greatly reduced if the Ni-MH battery is on occasions completely discharged before recharging. The usual recommendation is that this should be done after 3–5 normal charge/discharge sequences.

Table 2.3 shows the advantages and disadvantages of using batteries as power sources for electronic equipment, as compared to mains supplies.

Table 2.3   Battery operated equipment

Advantages	Disadvantages
Equipment is portable.	Limited energy capacity.
An ordinary battery is smaller and lighter than is any form of connection to the main supply.	Voltage generally low.
Safer to use, no high voltages being involved.	Batteries deteriorate during storage.
Equipment requires no trailing leads.	Batteries for high voltage or high current operation are heavier and more bulky than the equivalent mains equipment.

Question 2.5

Give two reasons for using nicad cells in portable tools rather than other types such as zinc or alkali manganese.

Lithium-ion cell

This rechargeable cell avoids the direct use of the metal lithium because it oxidizes too readily. A lithium-ion cell must never be broken open. The carbon anode is formed from a mixture of compounds at about 1100°C and then electrochemically doped with a lithium compound. The cathode is formed from a mixture of the compounds of lithium, cobalt, nickel and manganese. A mixture of propylene carbonate and diethyl carbonate (which avoids using water) is used for the electrolyte. This cell is nominally rated at 3.6 volts and has a long self discharge period, typically falling by 30% after 6 months. The recharge period is typically about 3 hours and the cell can withstand at least 1200 charge–recharge cycles.

In addition to its advantages with holding increased energy, the lithium-ion cell tends not to suffer from the memory effect, ensuring a longer life even when poorly treated. Its features include high energy density and high output voltage with good storage and cycle life. Lithium-ion cells are used in desktop personal computers (to back up memory), camcorders, cellular phones and also for portable CD players, laptop computers, PDA, and similar devices.

The cell operates on the principle that both charging and discharging actions cause lithium ions to transfer between the positive and negative electrodes. Unlike the action of other cells, the anode and cathode materials of the lithium ion cell remain unchanged through its life.

Battery charging

Lead-acid cells need to be recharged from a constant-voltage supply, so that when the cell is fully charged, its voltage is the same as that of the charger, and no more current passes. By contrast, nickel-cadmium cells must be charged at constant current, with the current switched off when the cell voltage reaches its maximum. Constant-current charging is needed so that excessive current cannot pass when the cell voltage is low.

Nickel-metal-hydride cells need a more complicated charger circuit, and one typical method charges at around 10% of the maximum rate, with the charging ended after a set time. Some types of cells include a temperature sensor that will open-circuit the cell when either charge or discharge currents cause excessive heating. For some applications trickle charging at 0.03% of maximum can be used for an indefinite period.

Lithium-ion cells can be charged at a slow rate using trickle-chargers intended for other cell types, but for rapid charging they require a specialized charger that carries out a cycle of charging according to the manufacturer's instructions.

Note that there is no such thing as a completely universal battery charger and though several microprocessor-controlled chargers are available they cannot be used on all types of cells or on mixed sets of cell types.

Solar cells

Although a number of materials can be used to convert energy from light (photoelectric energy), silicon is the most frequently used. Silicon solar cells are constructed from a matrix of silicon PN junctions. Each junction is typically five centimetres square and about half a millimetre thick, with the junction being formed very close to the upper surface. This is covered with a metallic grid to provide one contact and allow light to penetrate easily into the junction to generate the photo-voltaic potential. The back surface is covered with a metallic layer to provide a second low resistance connection.

Each cell is capable of generating a maximum open circuit voltage of 0.6 V, with a typical operating value of 0.4 V. Such a cell will support a short circuit current of about 100 mA. Solar generators are constructed from a series parallel arrangement of individual cells and a panel of 1 square metre can easily provide upwards of 100 watts of power. Note for comparison that a steam generator with a grate area of 1 square metre would be able to generate several hundred kilowatts – this is an example of the energy *density* factor that works in favour of conventional generators and against all alternative forms of energy conversions. Table 2.4 compares primary and secondary cells.

Table 2.4   Primary and secondary cells compared

Primary cells	Secondary cells
Low cost	Expensive
Small size	Comparatively long life
Short life	Rechargeable
Throwaway when exhausted	Generally heavier than equivalent primary cell
Light weight	Specialized products, less easily obtainable
Readily available	

Large value capacitors

Capacitors with a value of about 1 to 5 farads which can be charged to 5 volts through a high value of resistance, can support a small (backup) discharge current for many hours. Modern construction provides a device of only about 5 mm high with the same diameter, so that these can be used to provide a backup power supply for circuits that need only a small current to maintain operation throughout short duration power failures.

Conductors, insulators and resistance

Conductors are materials that allow electric current to flow through them, and which can therefore form part of a circuit in which a steady current flows. All *metals* are good conductors. Gases at low pressure (as in neon tubes) and solutions of salts, acids or alkalis in water will also conduct electric current.

Insulators are materials that do not allow a steady electric current to flow through them and they are therefore used to prevent such a flow. Most of the insulators that we use are solid materials that are not metals. Natural insulators, such as sulphur and pitch, are no longer used; and plastic materials like polystyrene and polythene have taken their place. Pure water is an insulator, but any trace of impurity will allow water to conduct some current, so that this provides one way of measuring water purity.

A good example of the contrasting uses of insulators and conductors is provided by printed circuit boards. The boards are made of an insulator, typically stiff bonded paper called 'SRBP' which is impregnated with a plastic resin (the name 'SRBP' stands for 'synthetic-resin bonded paper'); but the conducting tracks on the boards are made of conducting copper or from metallic inks. Boards made from fibreglass are used for more demanding purposes.

Both *insulation* and *conduction* are relative terms. A conductor which can pass very small currents may not conduct nearly well enough to be used with large currents. An insulator which is sufficient for the low voltage of a torch cell would be dangerously unsafe if it were used for the voltage of the mains (line) supply.

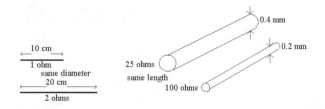
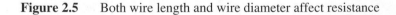

Figure 2.5   Both wire length and wire diameter affect resistance

The amount of conduction or insulation of materials is measured by their *resistance*. A very low resistance means that the material is a conductor; a very high resistance means that the material is an insulator. The resistance of any sample of a substance measures the amount of opposition it presents to the flow of an electric current. A long strip of the material has more resistance than a short strip of the same material. A wide sample has less resistance than a narrow sample of the same length and same material. These effects are illustrated in Figure 2.5.

The resistance of a sample of a given substance therefore depends both on its dimensions and on the material itself. As a formula, this is

$R \propto L/A$

where L represents the length of the sample and A its area of cross-section. The sign $\propto$ means 'is proportional to'. The formula therefore reads in full as: 'Resistance is proportional to sample length divided by the sample's area of cross-section'.

Resistance is measured in units called **ohms** (symbol Ω), which will be defined shortly. We can use the formula at present by using proportion.

Example: A wire 2 metres long has a resistance of 5 ohms. What would be the resistance of a 1.5 m sample of the same wire?

Solution: If a 2 m sample has a resistance of 5 ohms, the resistance of a 1 m sample must be 5/2 = 2.5 ohms. A 1.5 metre length will therefore have a resistance of $2.5 \times 1.5 = 3.75$ ohms.

Example: A sample of wire of radius 0.2 mm has a resistance of 12 ohms. What would be the resistance of a sample of the same material of the same length but having a 0.3 mm radius?

Solution: The area of a cross-section is proportional to the square of the radius. Therefore Resistance $A \times$ (Radius of A)^2 = Resistance $B \times$ (Radius of B)^2. Substitute the data, and simplify to get $12 \times 0.2^2 = R \times 0.3^2$

The resistance value of a sample of material does not depend only on its dimensions, however, because every material has a different value of resistance per standard sample.

Question 2.6

A sample of wire has a resistance of 2 ohms for a length of 36 metres. What resistance value would you expect for a length of 22 metres of the same wire?

This resistance per standard sample is called the *resistivity* of the material in question. It is measured in units called ohm-metres (written as Ωm). The resistance of a sample of any material is then given by the formula:

$$R = \frac{\rho \times L}{A}$$

where ρ (the Greek letter called 'rho') represents the *resistivity* of the material.

Example: Copper has value of resistivity equal to 1.7×10^{-8} Ωm. What is the resistance of 12 m of copper wire of 0.3 mm radius?

Solution: Substitute the data in the equation $R = \rho L/A$, putting all lengths into metres and recalling that $A = \pi r^2$, Then

$$R = \frac{1.7 \times 10^{-8} \times 12}{\pi \times (0.0003)^2} = \frac{20.4 \times 10^{-8}}{2.83 \times 10^{-7}} = 0.72 \text{ ohms}$$

Material	Resistivity
Aluminium	2.7
Copper	1.7
Iron	10.5
Mercury	96
Manganin	43
Bakelite	10^5
Glass	10^{12}
Quartz	10^{20}
PTFE	10^{20}

The resistivity values of some common materials are listed in the table, left. The resistivity figures need to be multiplied by 10^8 to give values in ohm-metres.

The material listed as *manganin* is a copper-based alloy containing manganese and nickel, much used in the construction of wirewound resistors. PTFE is the high resistivity plastic material whose full name is polytetrafluoroethylene.

Copper has a very low value of resistivity, so that copper is a good conductor and it used in electrical cables. Of all the metals, only silver has lower resistivity, but silver is too expensive to use for cables, though it is often used for small conductors. Aluminium is used for high-tension cables because an aluminium cable that has the same resistance as a copper cable is thicker but has less weight.

Activity 2.2

Use a resistance meter and (if available) a Megger, to measure the resistance of a metre of copper wire, a metre of Nichrome wire, and a square of paper.

Answers to questions

2.1   The heat loss is proportional to I^2, so that high voltage, low current results in lower heat loss.

2.2   0.005 A or 5×10^{-3} A, 0.000 046 A or 4.6×10^{-5} A, 0.000 000 221 A or 2.21×10^{-7} A.

2.3   35.66.

2.4   114 bhp.

2.5   Cost, current output.

2.6   1.22 ohms.

3

Resistors, networks and measurements

Resistors

20%	10%	5%
1.0	1.0	1.0
		1.1
	1.2	1.2
		1.3
1.5	1.5	1.5
		1.6
	1.8	1.8
		2.0
2.2	2.2	2.2
		2.4
	2.7	2.7
		3.0
3.3	3.3	3.3
		3.6
	3.9	3.9
		4.3
4.7	4.7	4.7
		5.1
	5.6	5.6
		6.2
6.8	6.8	6.8
		7.5
	8.2	8.2
		9.1

Resistors are components that have a stated fixed value of resistance. Methods of construction include wirewound, carbon moulded, carbon film and metal film resistors. Two symbols for a resistor in a circuit are shown, left. The older zigzag symbol is still used, particularly in the USA and Japan, and the rectangular symbol is used in Europe, but both are British Standard symbols.

Wirewound resistors are constructed from lengths of insulated wire, often a nickel chromium alloy, and they are used when comparatively low values of resistance are required. They are particularly useful when the resistor may become very hot in use, or when precise resistance values are needed.

Moulded carbon resistors are cheap to manufacture but cannot be made to precise values of resistance. Their large tolerances mean that the values of individual resistors have to be measured and suitable ones selected if a precise value of resistance is needed.

Carbon and metal film resistors are made, as the name suggests, from thin films of conducting material. They can be manufactured in quantity batches, but to fairly precise values. Moulded resistors are no longer in use in the UK, though they can be found on some imported equipment.

Values of resistance are given in ohms (Ω). In circuit diagrams, they are coded R for 'ohms', K (officially) or k for kilohms (thousands of ohms), and M for megohms, or millions of ohms. The letters R, K, k, or M are placed in the position of the decimal point when a resistance value is written against the resistor symbol, as was explained in Chapter 2.

Resistors are manufactured with certain average values aimed at. These values are called *preferred values*. They are chosen so that no resistor, whatever its actual value of resistance, can possibly lie outside the range of all tolerances. The table, left, lists the 20%, 10%, and 5% tolerance preferred values.

These figures are used for all resistor values of a given tolerance range. Values of 2R2, 3K3, 47K, 100K, 1M5 could thus lie in either range; but values of 1R2, 180R, 2K7, 39K, 560K could lie only in the 10 per cent (or closer) tolerance range.

Take, for example, a 2K2 resistor in the 20 per cent series. It could have a value of 2K2 ± 20% that is to say, a range from 2K64 to 1K76. But if such a resistor has a measured value of 2K7 or 1K6, it would not become a reject. The reason is that the next larger preferred value is 3K3 and 3K3 − 20% is 2K64. So the 2K7 resistor would quite legitimately be reclassified as a 3K3. Similarly on the low side, the next value down from 2K2 is 1K5, and 1K5 + 20% is 1K8. A resistor with a measured value of 1K6 would therefore be acceptable as a 1K8.

In other words, the preferred values are so chosen that no resistor that is made can be rejected simply because of its measured value. The preferred value of a resistor is either printed on the resistor or else coded on it by reference to the colour code illustrated and explained below.

E Series resistors

Within Europe, resistors are also described by the letter E and a digit that indicates the number of resistors per decade and this is related to the value tolerance as follows; 20% – E6, 10% – E12, 5% – E24 and 1% – E96. To accommodate the greater number of components per decade, E96 devices are coded by using 5 or 6 coloured bands.

Colour codes

The basic relationship between values and colours is as follows:

Black	Brown	Red	Orange	Yellow	Green	Blue	Violet	Grey	White
0	1	2	3	4	5	6	7	8	9

One of the many ancient mnemonics designed to aid memorizing this series goes, 'Bye Bye Rosie Off You Go Birmingham Via Great Western'.

Question 3.1

A 10% tolerance resistor is measured and the indicated value is 4K1. What is its most likely preferred value?

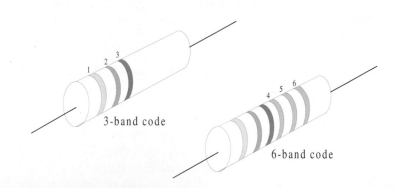

Figure 3.1   Resistor colour coding

In the 3-band code resistor pictured in Figure 3.1, all the colour bands are printed towards one of its ends. The first band (the one nearest the end itself) represents the first figure of the coded value; the second band represents the second figure; and the third band the number of zeros following this second figure. The three bands painted in the sequence, beginning from the end of the resistor – blue, grey, red – would mean a resistor with a preferred value of 6–8–two zeros, or 6800 ohms. Such a resistor would in practice always be described as a 6K8.

This colour coding system can be extended by adding a fourth band to indicate the component tolerance. No fourth band means ± 20%, but a gold band indicates ±5% and a silver band indicates ±10% respectively. Up to six bands may be used for close-tolerance resistors as shown in the 6-band example of Figure 3.1. Here the first three bands are used to represent significant figures whilst the fourth band indicates the multiplier or number of following zeros. Bands five and six then indicate component tolerance and temperature coefficient respectively.

Moulded carbon resistors (now obsolete) are usually colour coded. Wirewound and film type resistors generally have the resistance value printed on them, using the R, K, M notation.

When a resistor is being selected, you have to consider its power dissipation as well as its resistance. A resistor that is rated at ¼ W runs noticeably hot when it is asked to dissipate power of ¼ W; and it will be damaged if it is expected to dissipate more power than that. In electronic circuits, most resistors have to dissipate considerably less than ¼ W, so that ¼ W or even ⅛ W types can be used. Component lists therefore specify only those few resistors that need higher ratings. The description: 2K2 WW 5W, for example, means 2200 ohm resistance, wirewound, 5 W dissipation. For calculations of how power is dissipated in watts see later in this chapter.

In addition, some resistors appear on circuit diagrams with safety warnings, meaning that if this resistor fails it must be replaced by another that is exactly of the same type. This is important, and a service engineer could be held responsible for any damage that was caused if an ordinary resistor was used to replace a special component.

When resistors fail, they commonly become open-circuit (o/c), rarely short-circuit (s/c) or they may change their value. This change of value may be either upwards or downwards, and a faulty resistor that has changed value is said to have gone high or gone low.

Temperature coefficient

Metals and other common conductors have positive values of temperature coefficient. Poor conductors and semiconductors often have negative values.

The effect of a change of temperature on a resistor is to change its value of resistance. This change is caused by alteration of the resistivity value of the resistor material rather than by the very small change of dimensions (length and cross-sectional area) that also takes place.

This effect of temperature on resistance is measured by what is called the *temperature coefficient of resistance*, which is defined as the fractional change of resistance per degree of temperature change.

Temperature coefficient values may be either positive or negative. A positive temperature coefficient means that the resistance value of a resistor at high temperature is greater than its resistance value at low temperatures. A negative temperature coefficient means that the resistance value at high temperatures is lower than its resistance value at low temperatures.

For example, if a resistor is quoted as having a temperature coefficient of +250 ppm/°C, this means a change of 250 units for each million units of resistance for each degree Celsius of temperature rise.

For a 100 K resistor raised in temperature from 20°C to 140°C (a change of 120°C), this means:

$$\frac{250}{1\,000\,000} \times 100\,000 \times 120 = 3000 \text{ ohms, or 3K0}$$

making the value a total of 103 K at the higher temperature.

Taking another example, if an insulator has a resistance of 100 M at 20°C and has a temperature coefficient of −500 ppm/°C, then at 200°C (a rise of 180°C) its resistance will drop by:

$$\frac{500}{1\,000\,000} \times 100\,000\,000 \times 180 = 9\,000\,000 \text{ or 9 M}$$

so that the resistance at 200° is 91 M.

Materials that are widely used in resistor manufacture normally have positive coefficients of small value. Insulating materials, on the other hand, generally have negative temperature coefficients, and insulation becomes less effective at high temperature.

Resistors in series and in parallel

Resistors can be connected either in series or in parallel, or in any combination of series and parallel. When resistors are connected in series, see Figure 3.2(a), the same current must flow through each resistor in turn. The total resistance encountered by the current is the sum of all their separate values.

When resistors are connected in parallel, Figure 3.2(b), the current flow is divided among the resistors according to their values, with the most current flowing through the lowest value resistor(s). The same voltage, however, is maintained across all the resistors, and the total current is equal to the sum of the currents in each resistor.

Any combination of resistors connected in series and in parallel is capable of being replaced, as far as the flow of current is concerned, by a single resistance of equivalent value. These single resistances are then combined by using the formulae given in Figure 3.2, until a single resistance equivalent in value to the resistance of the combination as a whole is achieved. This is called the *equivalent resistance*, and once an equivalent resistance value has been calculated, the effect of a set of resistors can be assessed.

The equivalent value of a set of resistors in series is simply the sum of the resistance values in the series. For example, the equivalent value of 6K8 + 3K3 + 2K2 connected in series is 6.8 + 3.3 + 2.2 = 12.3K, or 12K3. The equivalent resistance of the series is thus greater than is the value of any of the single resistors.

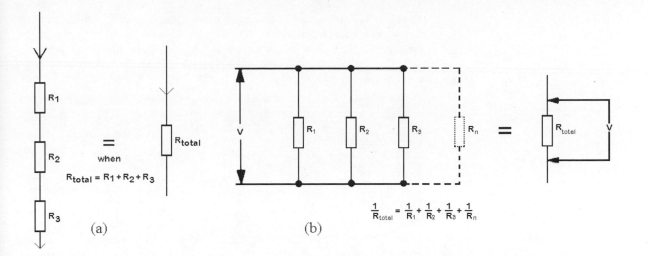

Figure 3.2   Connecting resistors (a) in series and (b) in parallel

The equivalent value for resistors connected in parallel is rather more difficult to calculate. The rule is to calculate the inverse (1/R) of the value of every resistor connected in parallel, to add all the inverses together, and then to invert the result. As a formula this is

For resistors in parallel, add the inverses of the values, and take the inverse of this sum

$$1/R_{total} = 1/R_1 + 1/R_2 + 1/R_3 + \ldots$$

– for however many resistors there may be in the parallel combination. For two resistors this can be simplified to Product/Sum, so that for two resistors R_1 and R_2, the equivalent is $R_1R_2/(R_1+R_2)$.

Example: Find the equivalent resistance of 2K2, 3K3, 6K8 connected in parallel.

Solution: $\dfrac{1}{R} = \dfrac{1}{2.2} + \dfrac{1}{3.3} + \dfrac{1}{6.8} = 0.4545 + 0.3030 + 0.1470 = 0.9046$

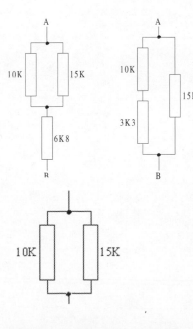

so that if 1/R is 0.9046, then R = 1/0.9046, which is 1.1054 K, or 1105.4 ohms. A calculator will invert a number when you press the Inv or X^{-1} key.

When a circuit contains resistors that are connected both in series and in parallel, the calculations must be carried out in sequence. The example, left, shows two circuits whose total equivalent resistance needs to be found. The procedure is set out, following.

In the example (a), the value of the parallel resistors is calculated first. The 10 K and the 15 K combine as follows

$$\frac{1}{R} = \frac{1}{10} + \frac{1}{15} = 0.166$$

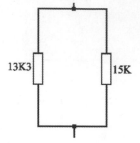

so that R = 6K (units of K have been used throughout). This 6K is now added to the 6K8 series connected resistor to give a total equivalent resistance of 12K8, which is the total resistance of the circuit. Note that we need to find the equivalent of the parallel resistors before we can add in the value of the series resistance.

In example (b), the series resistance values are combined first. The 10K plus the 3K3 gives 13K3, and this value is now combined in parallel with 15K.

$$\frac{1}{R} = \frac{1}{13.3} + \frac{1}{15} = 0.1418$$

so that R = 7.05K or 7K05

- Both series and parallel combinations of resistors can produce values of total resistance that are unobtainable within the normal series of preferred values. This can often be useful as a way of obtaining unusual values.

Question 3.2

What is the total resistance of a 5K6 in parallel with a 8K2 resistor?

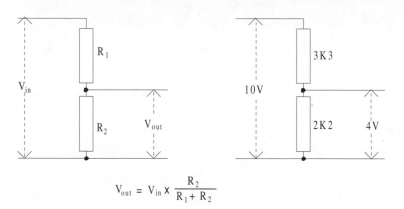

$$V_{out} = V_{in} \times \frac{R_2}{R_1 + R_2}$$

Figure 3.3   The potential divider circuit

A highly important circuit using two resistors connected in series is the *potential divider* circuit illustrated in Figure 3.3. The total resistance is $R_1 + R_2$ which in the example shown is 5K5. By Ohm's law, the current flow must be

$$\frac{V}{R_1 + R_2} = \frac{10}{5.5} = 1.8182 \text{ mA}$$

Such a current flowing through R_2 requires voltage V which, by Ohm's law again, must be $I \times R_2$ Substitute the known values of current and resistance, and the voltage across R_2 works out at 1.8182 mA $\times$ 2K2, or about 4 V in the example shown.

A potential divider (or attenuator) circuit is used to obtain a lower voltage from a source of high voltage. Where E is the higher voltage and V the lower voltage, the lower voltage can be calculated by using the formula:

$$V = \frac{ER_2}{R_1 + R_2}$$

For example, if 10 V is applied across a series circuit of 2K and 3K resistors, the voltage across the 3K resistor is:

$$V = \frac{ER_2}{R_1 + R_2}$$

$$\frac{10 \times 3}{3 + 2} = 6 \text{ V}$$

Example: Given the circuit shown in Figure 3.4(a), calculate V_{out}, I_1, I_2 and I_3.

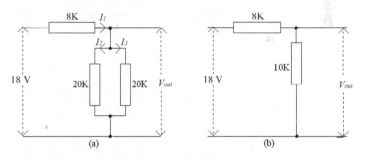

(a)   (b)

Figure 3.4   A further series parallel problem

Solution: Start with the parallel, equal 20K resistors. Use either the product/sum rule or recognize that two equal parallel resistors have a value equal to half that of one, so that the circuit becomes that shown in Figure 3.4(b). The total resistance is therefore 10K + 8K = 18K. Now 18 volts across 18K will cause a total current $I = 1$ mA. This current through 10K will produce a $V_{out} = 10$ volts. This is the voltage across both 20K resistors, so that I_1 and I_2 are both equal to 10/20K = 0.5 mA.

Potentiometers

Figure 3.5 shows the working principle of the component called the *potentiometer*, which is in effect a variable potential divider.

A sector of a circle, or of a straight strip, of resistive material is connected at each end to fixed terminals, and a sliding contact is held against the resistive material by a spring leaf. As the position of the sliding contact is altered, so will different values of resistance be created between the sliding contact and each of the two fixed terminals. The symbol illustrates the principle.

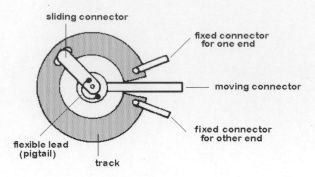

Figure 3.5   Circular potentiometer construction:

A variety of potentiometer circuit using a switch with separate fixed resistors is the switched potentiometer, the principle of which is shown in the drawing, left. The switch contact can be moved to connect to any of the points where resistors join, so that a number of potential divider circuits can be formed. There will clearly be a different value of voltage division at each contact point.

All resistors, whether fixed or variable, are specified by reference to their resistance value, their method of construction and their power rating. The abbreviation 'ww' or WW is normally used to denote wirewound resistors.

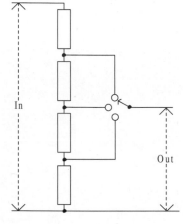

Joysticks

A joystick is used for controlling some types of action, and is used to a large extent along with a computer for games. The joystick can be moved on two dimensions, and the mechanical action operates two potentiometers that are set at right angles. Movement of the joystick forward and backward affects one potentiometer only, and movement left and right affects the other potentiometer. All other movements of the joystick will affect both potentiometers, so that the two potentiometer outputs provide signals that accurately represent the position of the joystick.

Measuring instruments

Electrical quantities are measured by instruments such as multimeters and oscilloscopes. A multimeter (or multi-range meter) is an instrument capable of indicating the values of several ranges of currents, voltages and resistances by measuring the current flowing through the meter.

To measure a voltage, the multimeter is first switched to the correct range of voltage and the leads of the multimeter are clipped to the points across which the voltage is to be measured. One of these points will often be either the chassis, earth or the negative supply line. The circuit is now switched on. The voltage is then read from the correct scale. See Figure 3.6.

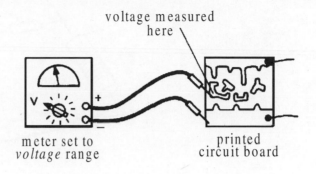

Figure 3.6   Measuring voltage on a printed circuit board

To read current flow, Figure 3.7, the circuit is broken at the place where the current is to be measured. The multimeter leads are then clipped on, one to each side of the circuit break. The meter is set to a suitable range of current flows, and the circuit is switched on. Current value is then read from the correct scale.

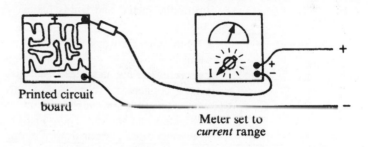

Figure 3.7   Measuring current flow on a printed circuit board

Note that, for measurements of either voltage or current, the meter must be connected in the correct polarity, with the + (usually red) lead of the meter connected to a more positive voltage.

To read resistance values, Figure 3.8, the multimeter leads are first short-circuited. The meter is switched to the 'ohms' range, and the set zero adjuster is used to locate the needle pointer over the zero ohms mark. The leads are then reconnected across the resistor to be measured, and its resistance is read off the scale.

Note that you should always check the zero adjustment when you make a resistance reading, because the zero reading will drift as the battery within the meter discharges. Do not rely on such readings for precise measurements because the scale is non-linear, so that, for example, a value midway on the scale between 2 and 3 is not 2.5.

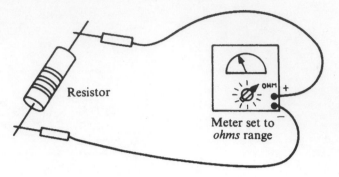

Figure 3.8   Measuring resistance

If there is any doubt as to what sort of values of voltage or current will be encountered, the highest likely range of each must first be tried on the multimeter. The range switch is then used to select a lower range of values, until a range is found which gives a reading which is not too near either end of the scale. Note that the resistance ranges make use of a battery inside the meter, and if this battery is dead resistance readings cannot be made.

Multimeters always draw current from the circuit under test, so the values of voltage or current which are measured when the meter is connected are not necessarily the same as those which exist when the meter is not connected. To overcome this problem (which is caused by the comparatively low resistance of the multimeter), electronic digital voltmeters can be used. These instruments have very high resistance and so draw very little current from a circuit. They should always be used when measurements have to be taken on high resistance circuits.

Activity 3.1

Connect the circuit shown in Figure 3.9. With the multimeter switched to the 10 V scale, connect the negative lead to the negative line. Measure the supply voltage, and adjust it to exactly 9 V. Measure the voltages at points X and Y, and record the two values.

Now use the potential divider formula to calculate what voltages would be present at X and at Y if no meter were connected. If an electronic voltmeter is available, use it to read the voltage at point Y. Which type of meter is preferable for measuring the voltage at Y, and why?

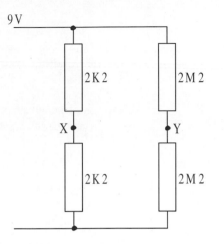

Figure 3.9   Circuit for Activity 3.1

Examples: These examples should be solved by mental arithmetic, but you can use a calculator to check your answers if necessary. In each question, two quantities are given and you should calculate the third.

V = 9 V	V = 6.6 V	V = ?	V = 27 V	V = 12 V	V = ?
R = 18K	R = ?	R = 10K	R = 1K8	R = ?	R = 2K2
I = ?	I = 2 mA	I = 3 mA	I = ?	I = 4 mA	I = 3 mA

Insulation testers are used to measure the very high resistances of insulators. Testers such as the well-known Megger operate by generating a known high voltage and applying it through a current meter to the insulator under test. Any current flowing through the insulator causes the meter needle to deflect, and the scale indicates the resistance value of the insulator in megohms. Typical proof test voltages are 250 V or 500 V d.c. and 500 V a.c. Megger is a registered trade name, but it is widely used to mean any meter for high resistance measurements.

The *cathode ray oscilloscope* uses a cathode ray tube to trace out the waveform of a signal input. A measuring graticule (taking in this case, the form of a transparent scale) enables the dimensions of the signal waveform to be measured. These dimensions can then be converted into voltage and time units by using the conversions marked on the range switches. In use, the CRO is connected like a voltmeter.

Signal generators are used to supply signal waveforms, usually sine or square waves at various frequencies. AF (audio frequency) signal generators have outputs whose frequency range is from about 10 Hz to 20 kHz, usually possessing enough power to drive a small loudspeaker. RF (radio frequency) signal generators work at high frequencies, typically 100 kHz to many MHz, and have both low amplitude and low power.

Measuring instruments such as these are used to test components and circuits. The multimeter can test resistance values and check that the correct currents and voltage levels exist in a circuit. Electronic voltmeters are used to test high resistance circuits. The CRO is an extremely useful instrument which enables the engineer to see the shape, as well as to measure the amplitude and period, of a waveform. In circuits which amplify signals, signal generators provide a signal input which can then be traced through the circuit using the CRO.

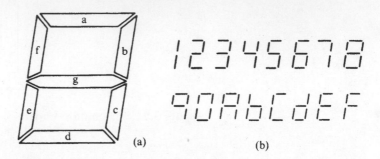

(a)   (b)

Figure 3.10   The seven-segment display

A *seven-segment* display is now much used in digital electronic instruments to display readings of voltage, current or resistance directly in figures. The basic display consists of seven segments arranged in the form of a slanting figure '*8*' (see Figure 3.10(a)). Each of the segments a, b, c, d, e, f and g, can be lit up to distinguish it from its background, so that the figures 0 to 9 (as well as a few letters of the alphabet) can be displayed by illuminating different segments (Figure 3.10(b)).

Seven-segment readouts need to be driven by a decoder circuit (generally taking the form of an integrated circuit) so that the correct segments are illuminated from the appropriate number input to the decoder. The display and decoder are very often combined, particularly when the LCD type of display is used.

Activity 3.2

Measure the resistance of a selection of resistors, and compare your readings with the marked values. Use either a multimeter or a resistance bridge for measurement.

Activity 3.3

Measure the insulation resistance of unpopulated PCBs and cables. Use a Megger or equivalent on the 1000M range, and make sure that a selected PCB is clean and dry. Place the PCB (or stripboard) on an insulator, such as polystyrene foam, and connect the two leads from the Megger to different tracks on the PCB or stripboard. Switch on the Megger and read the insulation resistance.

Activity 3.4

Now coil a length of television coaxial cable on to the polystyrene foam, and make certain that the inner and output conductors are well separated at the ends of the cable. Connect one of the Megger leads to the inner conductor of the cable and the other lead to the outer conductor. Switch on the Megger and measure the insulation resistance for the cable. Repeat this on a section of mains cable, and on a stereo amplifier connection cable.

Answers to questions

3.1   3K9 10%.
3.2   About 3K3.

4 Heat topics

Heat and temperature

The quantity called *temperature* measures the *level* of heat just as voltage measures electrical level (potential). Heat energy will flow from a place at a high temperature to one at a low temperature, just as naturally as water flows downhill or electric current flows from a point at high voltage to one at low voltage. Heat *never* flows in the reverse direction, from cold to hot, and it is this fact that causes so much heat energy to be wasted.

Temperature is, for most practical purposes, measured on the Celsius (*not* Centigrade) scale, in which 0°C represents the freezing point of pure water at sea level and 100°C its boiling point under the same conditions. Some electronic formulae (for thermistor action) require temperatures to be expressed in the absolute (or Kelvin) scale. A temperature of 0 K (no degree sign is used) represents *absolute zero*. No lower temperature is conceivable, either in theory or in practice. On the Kelvin scale, the freezing point of water is 273 K, and a comfortable 20°C room temperature is 293 K. To convert Celsius degrees into Kelvin degrees, add 273 to the figure of Celsius temperature.

Temperature is measured by the effects it has on materials.

Gases, liquids and solids expand when temperature is increased.

The resistance of metals increases as temperature is increased.

The resistance of insulators decreases as temperature is increased.

The voltage across different metals in contact with one another (contact potential) changes as the temperature is increased.

Expansion and expansivity

Changing the temperature of a solid material will cause a change of dimensions, but this change is very small. For example, if a steel bar has a length at 0°C of exactly 100 mm, then its length at a temperature of 100°C is 100.12 mm, not a change that is easy to observe. Liquids expand to a much greater extent. The quantity that we can use to measure expansion of a liquid is its volume, and one of the remarkable features of water is that its expansion depends on temperature range. As water is heated from 0°C to 4°C, its volume decreases, but from 4°C upwards, the volume increases.

Other liquids have more uniform expansion. For example, a volume of methanol (methyl alcohol) that is 100 ml at 0°C will become 106 ml at 50°. Different liquids have widely differing expansion characteristics, with water expanding very much less than solvents like methanol or benzene. The familiar mercury thermometer uses a thin glass bulb containing mercury which can expand into a narrow-bore tube. Because the tube has a narrow bore, a small change in the volume of mercury will cause a large change in the length of the mercury column in the tube.

The volume of gases is affected much more by temperature. If a gas is free to expand, maintained at a constant pressure, then its volume will change as the temperature changes. If the gas is confined in a constant volume container then its pressure will change as the temperature changes. The amount of change of volume or pressure is much the same for all gases, amounting to 1/273 of original volume (or pressure) per degree. For example, if a gas has a volume of 100 ml at 0°C and is at a constant pressure, heating it to 100° will change its volume by about 36 ml. Thermometers based on gas expansion can be much more sensitive to temperature change than those based on solids or liquids.

Thermometers and thermostats

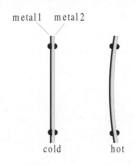

metal1   metal2

cold   hot

Mechanical thermometers and thermostats work by making the use of these expansion effects. Conventional thermometers use the expansion and contraction of mercury, and simple thermostats use the expansion and contraction of solid bimetal strips (see left). The expansion of mercury in a bulb forces the mercury level in a narrow-bore tube (a capillary tube) to rise, and the level will fall when the mercury contracts as the temperature is reduced. A bimetal strip bends as it is heated because one of the two metals of which it is composed expands, at a known rate, more than does the other. This is used in many types of electrical thermostats because the strip can be connected as a switch.

Resistance thermometers, are much more precise than the mercury type They make use of the effect of temperature change on the electrical resistance of metal wires, and thermocouples take advantage of the voltage generated because of contact potential changes when junctions of dissimilar metals are heated. Some semiconductor materials can also be used for sensing temperature because of their large change of resistance as temperature is changed. These are the materials used for thermistors.

Heat flow

Heat energy flows naturally from points at high temperature to points at a lower temperature. The three known methods of heat energy transfer are *conduction, convection* and *radiation*. All three have applications in electronics.

Heat is conducted through materials most easily through metals, hardly at all through gases, and poorly through liquids. We know that heat is conducted best by materials which possess large numbers of free electrons, in other words through materials which are also good electrical conductors. Heat can also be conducted by the movement of whole atoms, however; and some materials which are reasonably good conductors of heat are actually electrical insulators.

The importance of heat conduction in electronics is that this is the way in which all the heat generated in the inside of a transistor is dissipated to its case. The same heat is then often conducted even further away to a *heatsink*, Figure 4.1, a formed metal slab on which the transistor is mounted. A thin mica washer is often sandwiched between the transistor and the heatsink to provide electrical insulation while still allowing a reasonable amount of conduction of heat.

Heat-transfer greases are often used to reduce the thermal resistance between a transistor and its heatsink.

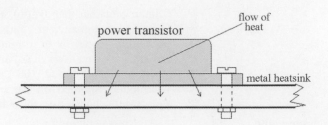

Figure 4.1   How heat is dissipated by conduction to a heatsink

Because heat flows only from a high temperature to a low temperature, the interior of a working transistor is always at a higher temperature than is its case or the heatsink itself.

Heat energy is transferred through gases by another action, called *convection*. A gas such as air is heated when it comes into contact with a hot surface. This heated mass of air expands, and this reduces its density (the mass per unit volume) compared to that of the cold air surrounding it. The hot air therefore rises (this is the principle of the hot-air balloon), carrying away heat with it, and allows cold air to take its place. This sets up a flow with air moving past the hot object and removing heat from it in the process. Convection is used to transfer the heat from transistors, resistors, heatsinks and other components to the air. Convection also applies to liquids.

The action of a heatsink is simply to present a larger surface to the air so as to make the removal of heat easier. Forced air cooling with the aid of fans greatly increases the rate at which heat can be removed, though water cooling is even more efficient.

Radiation has very little effect in removing heat from objects unless they are at a really high temperature. Radiation transfers energy by means of electromagnetic waves; and though all objects radiate heat to some extent, the quantity that is radiated becomes (compared to the losses by conduction and/or convection) significant only when the subject is at a temperature approaching red heat. Valves and cathode-ray tubes dissipate most of their heat by radiation. Solid-state electronic equipment is, by contrast, cooled mainly by convection and conduction.

Heatsink calculations

Heatsinks are a very important part of power supplies and of power output stages. In general, servicing is concerned with maintenance of existing heatsinks rather than replacement or construction, but some notes on heatsink theory and practice may be useful.

In all heatsink calculations, the thermal resistance is very important, and it is a particularly useful quantity because it so closely resembles electrical resistance, allowing us to imagine all the items in the flow path of heat as resistors in a circuit. The symbol used for thermal resistance is θ (Greek theta), and its definition is as shown here, left.

$$\theta = \frac{\text{temperature change in } °C}{\text{power transferred in watts}}$$

This emphasizes that when heat flows, there is always a temperature difference or temperature *gradient*, and the direction of heat flow is always from high temperature to lower temperature. The transfer of heat from the collector junction of a transistor (or from junctions within an IC) is by conduction, and for conductive heat transfer the thermal resistance of each part of the circuit depends only on the materials and the way that they are joined, not on temperature levels. Convective heat transfer through gases (including air) and liquids is not so simple, nor is radiative heat transfer across a gas or a vacuum.

The path from a collector junction, where electrical power is converted to heat, can be imagined as a set of thermal resistances in series, and, like electrical resistance in series, the quantities add to give a total. For example, if a transistor has a thermal resistance θ_{JC} from junction to case, and there is a thermal resistance of value θ_{CS} from the case to the heat sink, and θ_{SA} from the heatsink to the surrounding air (ambient), then the total thermal resistance in this 'circuit' is $\theta_{JC} + \theta_{CS} + \theta_{SA}$. Multiplying this quantity by the amount of power that is dissipated will give the temperature difference between the ambient air and the semiconductor junction, and adding the ambient temperature (often assumed as 50°C) to this gives the

$$T_J = T_A + P(\theta_{JC} + \theta_{CS} + \theta_{SA})$$

temperature of the semiconductor junction. The equation is illustrated, left. The junction temperature calculated from this equation must not exceed the maximum temperature stipulated by the manufacturer.

Very often, you need to find what the maximum value of thermal resistance θ_{SA} can be used, given the assumed ambient temperature and maximum semiconductor junction temperature, with suitable values of the other thermal resistances. In such a case, the equation can be re-cast as

$$\theta_{SA} = \frac{T_J - T_A}{P} - \theta_{JC} - \theta_{CS}$$

shown here. The θ_{JC} value can be obtained from the specification of the semiconductor, and case to heatsink value, using a mica washer and heat-sink grease is usually of the order of 0.50°C/W.

For example, suppose that the junction temperature is not to exceed 90°C for a power dissipation of 10 W in air at 50°C. The thermal resistance from junction to case is specified by the manufacturer as 1.5°C/W, and the mica washer and grease contributes 0.50°C/W. The maximum possible heatsink thermal resistance is then given by entering the values into the formula, so

$$\theta_{SA} = \frac{90 - 50}{10} - 1.5 - 0.5 = 2$$

that the heatsink must have a thermal resistance of less than 2°C/W. Note that if your calculation results in a value that is zero or negative, then no heatsink can possibly be good enough, and the power level must be reduced, or a better heatsink used.

NOTE: The heatsinks used for computer microprocessor chips incorporate a fan for cooling, and their contact surface is pre-coated with a form of heat-transfer compound that melts into place when the processor is initially powered up. On a modern computer, some other chips will also be cooled in this way, and the chips incorporate temperature measuring circuits that will shut down the power supply if the chip temperature rises excessively.

Question 4.1

If heat flows from hot to cold, how do refrigerators and air-conditioners work?

Fuses

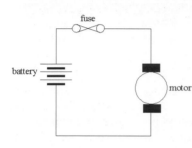

A fuse is made using a thin wire made from a metal with a fairly low melting temperature. The wire is usually surrounded by a heat insulator, so that the heat which is generated when a current flows through the wire is not quickly dissipated. When excessive current flows, the temperature of the wire rises rapidly until the wire melts and so breaks the circuit. The drawing, left, shows the symbol for a fuse in a simple circuit.

If a fuse is used in a circuit in which a high voltage would exist after the fuse blows, then that voltage placed across the fuse might be enough to cause an arc-over, so that current was not interrupted when the fuse blew. Fuses carry a voltage rating, and should not be used beyond that rating, so that a fuse rated at 125 V should not be used in a circuit in which 240 V could exist across a blown fuse. A fuse can be safely used at voltages lower than the rated maximum, but not at any voltage higher than the stated maximum.

The quantity that is always quoted for a fuse is its nominal current rating, but precisely what that means depends on the standard to which the fuse has been constructed. A fuse which is rated at 1 A, for example, will not necessarily blow at a current of 1 A, because the blowing of a fuse is a complicated process that involves current and time, and the various standards exist to provide guidelines on the current–time limits.

Fuses are grouped in five main categories according to their current-time characteristics. At one end of the scale, semiconductor circuits need fuses that will act very quickly on short-circuit conditions. These are now described as super quick-acting fuses, coded FF. These, at 10 times rated current will blow in a millisecond or less, and for twice the rated current the blowing time will be 50 ms or less. The next group comprises the quick-acting class F that are used for general-purpose protection where current surges are unlikely to be encountered. These have a slower blowing characteristic, some 10 ms for ten times rated current and just over 100 ms for twice rated current.

The medium time-lag fuses, type M, will withstand small current overloads that might be caused by charging capacitors. These fuses will blow after about 30 ms on a ten-times current overload, and after about 20 seconds on a two-times overload. The time-lag type T fuses will blow in 100 ms for a ten times overload and in about 20 seconds for a twofold overload. Super time-lag class TT fuses allow for 150 ms at a tenfold overload and 100 seconds at a twofold overload.

All fuse ratings are measured at 20°–25° ambient temperature, and because the blowing of a fuse is a heating action, the ratings of a fuse are affected by changes in the ambient temperature.

Fuses are resistors, and though the larger capacity of fuses have a negligible resistance, this is not true of the smaller types. Fuse resistance is not usually quoted by suppliers, but it can add to the resistance of a power supply and upset stabilization to some extent, though for the larger rated fuses the contact between fuse and fuseholder contributes more resistance in some cases.

Question 4.2

What do you think would be the effect of wrapping thermal insulating material, such as glass-fibre, around a fuse?

Panel lamps

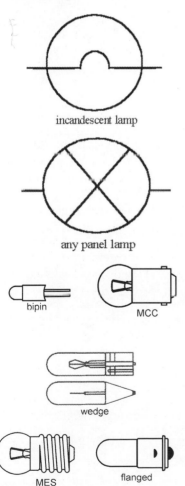

incandescent lamp

any panel lamp

bipin

MCC

wedge

MES

flanged

Panel lamps are usually the familiar incandescent (meaning hot-filament) type of lamp, usually in the MES (medium Edison screw), or the bayonet fitting MCC (Ba9s) or MBC fitting. Note that there is a vast range of different sizes and fittings, and bulbs that look similar are not necessarily a correct replacement. The BS symbols are illustrated here.

Panel lamps must be mounted in a holder that is secured to the panel, and coloured lens caps can be used to determine the colour of the indicator. Remember that a panel lamp is a resistor and that even small lamps need to dissipate their heat. The life of a panel lamp will be very short if its heat is not dissipated, and frequent failure (assuming correct supply voltage) can usually be cured by better cooling.

Panel lamps are classed by physical size, using letters such as M for miniature and L for Lilliput (smaller than miniature). They are further classed by the type of connector, with B meaning bayonet (push and twist) and S for screw. Some common types are listed in table 4.1. The typical values that are quoted apply to the most common types used for electronics applications. For other applications, such as on 'white goods' electrical products, panel lamps operating at voltages up to mains 240 V are used.

Table 4.1 Small panel lamps

Type	Typical voltage range, V	Typical current range mA	typical life (hours)
MES	1.25–3.5	200	10–1 000
MCC	6–24	50/100	1 000–3 000
flanged	12–28	100–400	5 000–25 000
Bi-pin	28	24	5 000
wedge	28	24	5 000

- LEDs are also used as panel indicators. They have a longer life than incandescent bulbs but are not usually so bright and with little choice of colour.

Answers to questions

4.1   Actions such as expanding a gas or passing current between metal junctions can cause cooling in one place and heating in another. Heat then travels to these cool regions. In a refrigerator, the heat is dissipated from a coil of pipes at the back.

4.2   The fuse will blow at a lower current rating and its blowing speed will be faster.

Unit 2

Outcomes

1. Demonstrate an understanding of electromagnetic devices and capacitors and apply this knowledge safely in a practical situation.

2. Demonstrate an understanding of alternating current and voltage and apply this knowledge safely in a practical situation.

3. Demonstrate an understanding of a.c. mains supply safety and distribution and apply this knowledge safely in a practical situation.

Note: Because it is more logical to treat alternating current ahead of electromagnetic devices and capacitors, Outcome 2 has been dealt with first.

5 Alternating currents

Electrical current can be *d.c.* or *a.c.*, and mixtures of both. D.c. has been featured in Unit 1, and it is a steady flow of current as is used to operate electronic equipment. A.c. means alternating current, meaning that the current is not steady; its value rises, falls, reverses, and rises and falls in the reverse direction. A.c. current is a form of wave, and when the voltage or current of mains a.c. is plotted on a graph it provides the shape of a sine wave. A.c. is generated by rotating machines such as alternators, and by electronic circuits called oscillators.

A.c. is used for electricity distribution because it is easy to change the voltage of a.c. The device that we call a transformer will convert one voltage of a.c. into another, so that we can convert a 240 V supply into a 6 V supply by using a *step-down* transformer.

When the voltage is transformed to a lower value, the amount of current that you can draw at the lower voltage is greater than the amount drawn at the higher voltage. For example, if you take a transformer that has an input (primary) of 1000 V a.c. and its output is 100 V, drawing 1 A, then the input current will be only 0.1 A. The voltage ratio is 10:1, and the current ratio is 1:10, the inverse. When a.c. voltage is transformed to a higher value, the current at the high voltage is lower than the current at the lower voltage. This is because theoretically all of the power developed in the primary winding is transferred to the secondary.

As a component, the transformer is indicated (see left) by a symbol of two coils, and this is in fact how a transformer is made, by winding two separate coils of wire around an iron core, see Chapter 6.

A.c. has to be used for electricity transmission because of the voltage loss and heat loss that occurs when a current flows through a cable. Because transmission cables are very long, we can keep losses to a minimum only if the currents in the cables are very low. We can use low currents if the voltages are very high, so that voltages of around 500 kV are used. These cables have to be mounted on high pylons with huge insulators so as to avoid sparking (arcing) to the ground or to each other. Where electricity is used, a transformer will convert the a.c. to a lower voltage, with more current available. This is why each housing or factory estate has its own transformer or set of transformers. The process is usually done in stages, with a transformer in each stage converting from high voltage to a lower value.

We can convert a.c. to d.c. fairly easily, and convert d.c. to a.c. using more complicated circuits called *inverters*.

Electronic signals are a.c., but not like the a.c. we use for power transmission. Signal voltages and currents are often very much lower, measured in millivolts and microamps (thousandths of a volt and millionths of an amp). In addition, the frequency of signals is often much higher than the 50 Hz used for power transmission, and the waveshape need not be a

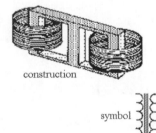

construction

symbol

sine wave. Signals can be of frequencies that need units such as kHz, MHz and GHz. Electronic circuits are used to work with signals, using d.c. to provide power.

Components called capacitors and inductors are used extensively in a.c. circuits. We shall learn more about these later, but in brief:

(a)   A capacitor blocks the flow of d.c. but provides a path for a.c.

(b)   An inductor provides a low-resistance path for d.c., but restricts the flow of a.c.

Waveforms

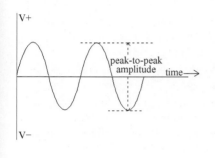

A waveform is most easily represented by a graph of either voltage or current plotted against time, as illustrated here, left. The same waveform should also be seen when a CRO is applied to the correct point in a circuit, and measurements of waveform quantities (or parameters) at that point can be made. The most useful measurements are generally those of wave *amplitude* and *time period* (or wave duration).

The amplitude of a waveform is the value of the voltage or current, and it varies during one cycle of a wave. The *peak-to-peak amplitude* of a wave is shown in the sketch, left, but is more easily measured on the face of a CRO tube (Figure 5.1) where it is found by measuring the vertical distance in centimetres between peaks of the wave and then multiplying this distance by the settings of the *Volts/cm* switch of the oscilloscope.

For example, a peak-to-peak (p–p) distance of 3 cm at a setting of 5 V/cm corresponds to 3×5 V = 15 V p–p. On a graph, the p–p amplitude can be read from the calibrated vertical scale of the graph.

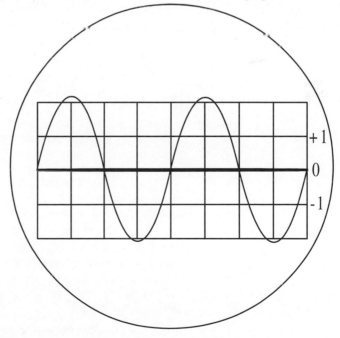

Figure 5.1   Peak-to-peak amplitude being measured on an oscilloscope

Peak amplitude, as opposed to *peak-to-peak amplitude*, is sometimes used as a measurement when the waveform is *symmetrical*. A symmetrical waveform has exactly the same shape above and below its centre line. Figure 5.2 shows a sine wave (which is always symmetrical) and a square wave (which is not always symmetrical, but is in this example).

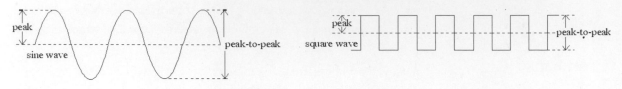

Figure 5.2   Peak amplitude and peak-to-peak amplitude on symmetrical waveforms

Meters that are used to read values of wave amplitude do not normally measure the peak-to-peak amplitude. A meter that is set on its d.c. voltage (or current) range will usually read zero when it is used for a.c. An a.c. meter is calibrated to read a quantity called root mean square (r.m.s.) value, see below. A waveform will have (Figure 5.3) an average value of current flow in one direction if the area of the graph of the wave above (or below) the centreline of the graph is greater than the area of the graph of the wave in the other direction. This would produce a reading on a d.c. meter as well as on an a.c. meter, and these reading amounts would not be the same. When a mixture of d.c. and a.c. is present in a circuit, the average value as measured by a d.c. meter shows the amount of d.c. present.

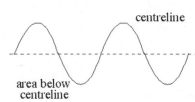
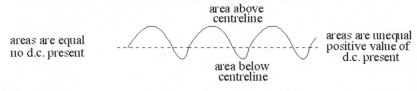

Figure 5.3   The d.c. component of differing waveforms

R.m.s.

Another method of measuring amplitude is used mainly for sine waves. It is called the *root mean square*, or *r.m.s.* When a sine wave has a value of current flow of 1 A r.m.s., it means that it will produce the same amount of power in a resistor as will a steady (d.c.) current of 1 A.

Equally, 1 V r.m.s. of a sine wave applied across a resistor will produce the same amount of power dissipation as will one volt of d.c. When a waveform of a.c. is a sine wave, the r.m.s. value of its amplitude will be 0.707 times its peak value (which is the same as the peak value divided by the square root of 2).

Waveforms of other shapes have different relationships between their r.m.s. and peak values; so meter readings of r.m.s. voltages and currents should be taken only when sine waves are being measured. Some specialized instruments exist which measure the r.m.s. value of *any* waveform; but you will not find them much used in servicing applications.

Activity 5.1

Use a signal generator connected to an oscilloscope to view and measure sine and square waveforms.

Question 5.1

A sine wave is measured as 14 V peak to peak. What does this correspond to in an r.m.s. meter reading?

R.m.s. quantities are important for power calculations, because they can be used for a.c. in exactly the same way as the d.c. voltage and current reading can be used for d.c. For example, the equations for power dissipated in a resistor R with current *I* flowing and voltage *V* across the resistor will be:

$$P = V.I \ \textbf{ or } \ P = V^2/R \ \textbf{ or } \ P = I^2R$$

Whether *V* and *I* are d.c. values or a.c. r.m.s. values.

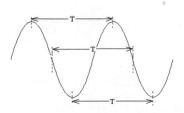

The periodic time (or duration) of a waveform is another important quantity which can be measured by using an oscilloscope, or which can be read from a graph. The *period* of a wave is the time taken for a complete cycle of the wave, as illustrated here.

To measure this quantity on the CRO display, the waveform must be *synchronized* or *locked* which means that the display must not be moving across the screen. The distance in cm between corresponding points is then measured. The corresponding points may be either the positive or the negative peaks, or the zero voltage levels at the points where the voltage is changing in the direction, or any other features which repeat once and once only, per cycle. The measured distance in cm is then multiplied by the setting of the 'Time/cm' switch to give the time period in seconds, ms or μs. Special care is needed when this reading is taken because some models of oscilloscope incorporate potentiometer control of time/cm, in addition to the switch and this potentiometer must be set to one end of its travel before the time reading is taken.

The frequency of a wave is the number of complete cycles of the wave accomplished per second and is equal to 1/Period. If the period is measured in seconds, 1/Period gives the frequency in units of hertz (Hz), or cycles per second. If the period is measured in milliseconds (ms), the frequency value will be in kilohertz (kHz). If the period is measured in microseconds (μs) the frequency value will be in Megahertz (MHz).

Frequencies

AF

IF

RF

VHF

SHF

The frequencies used in electronics are often identified by their range of value rather than by quoting exact frequencies. Thus the expression low frequency (LF) covers all frequencies below 50 Hz extending right down to d.c. *Audio frequency* (AF) covers the range of frequencies from about 40 Hz to 20 kHz which is about the range of sound wave frequencies which can be detected by the (younger) human ear.

Frequencies higher than the audio range are collectively known as *radio frequencies* (RF), but certain ranges of RF have distinctive names. For example, the range 460 kHz to 470 kHz is called *intermediate frequency* (IF) because of its special use in AM superhet receivers (see Chapter 16). Frequencies in the range 30 MHz to about 300 MHz are called very high frequency (VHF), and frequencies above about 300 MHz ultra high frequency (UHF). The UHF range is often taken as 300 MHz to 3 GHz (3 000 MHz), with the range 3 GHz to 30 GHz labelled as SHF (Super High Frequency, and frequencies above 30 GHz as Extremely High Frequencies (EHF). These names are misleading and difficult to remember, and it is always better to refer to the frequency range in MHz or GHz. More often now we use the term *microwave frequencies* to mean the frequencies from around 1 GHz upwards. In this range, we use the range of around 10 GHz to 13 GHz for satellite TV transmission, and Global Positioning Satellite systems (GPS) use a frequency around 1.5 GHz. Mobile phones use two bands of frequencies in the range 890 MHz to 960 MHz. Radar uses frequencies ranging from 15 GHz upwards.

One particular frequency, 2.45 GHz, resonates with the natural frequency of vibration of water molecules, and so this is the frequency used for microwave ovens. No other frequency has such a strong heating effect on water or materials containing water.

All these names are purely matters of convenience, lacking precise definitions. For example, frequencies of 100 kHz or so can generally be handled by the same circuits as are used for audio frequencies. Such frequencies could logically, therefore be classified as audio frequencies, even though a 100 kHz sound wave cannot be heard by any human ear. It is referred to as an *ultrasonic frequency* which means that it lies beyond the frequency range which the human ear is capable of detecting.

Power distribution uses very low frequencies, 50 Hz in Europe and 60 Hz in America and Japan. Conventional telephone lines can use frequencies in the range 100 Hz to around 3 kHz.

Though audio frequencies are the same for both sound and for audio-frequency radio waves, these are entirely different waves. One important difference is propagation method. Sound waves are mechanical waves that propagate in solids, liquids or gases by vibrating the molecules. Each vibrating molecule passes the vibration to its neighbour, so that the wave propagates at a speed that depends on the density and elasticity of the material. Sound cannot travel in a vacuum. The speed of sound is very much faster in liquids and dense solids than it is in air (speed around 300 metres per second), so that sound waves travel faster in liquids and

direction of vibration ← → direction of wave movement
longitudinal

direction of vibration ↕ direction of wave movement
transverse

solids. In addition, the wave is a longitudinal one, meaning that the molecules vibrate in the same line as the line of motion of the wave.

Radio waves at any frequency propagate because of the oscillation of electric and magnetic fields, and this can happen most easily in a vacuum (called free space). The waves are transverse, at right angles to the direction of propagation, and their speed is typically 300 million metres per second in free space, slower in other materials.

Graphs

Throughout this book, we shall use graphs to illustrate information about electronic components, systems and circuits. You need to understand both how to plot graphs and how to interpret the information they present. In addition, the most useful servicing tool is the *oscilloscope* which displays a form of graph on its screen.

The purpose of a graph is to take the place of a detailed table of measurements, so that it shows you at a glance how two quantities are related to one another. The table of voltages and currents in the table, left, for example, lists the results of a series of measurements which have been taken on a given circuit.

Try to guess quickly from this table alone the value of voltage that would result in a current flow of, say, 0.2 A. Then see how much quicker and more accurate your estimate would be if the same information is shown in the form of the graph which forms Figure 5.4.

V in volts	I in amps
0.85	0.07
1.9	0.15
3.0	0.24
3.9	0.31
5.4	0.43

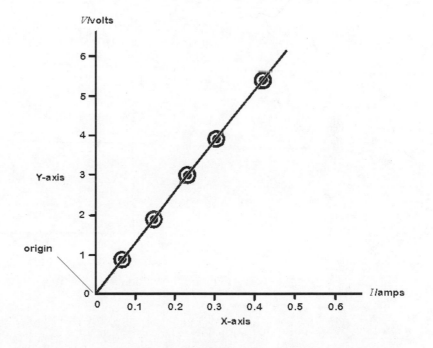

Figure 5.4   A typical linear graph

Y

X

In this graph, the relationship between the two sets of data is at once obvious, and the reasonably accurate values of all current flows and their corresponding voltages can be read from the graph scales at a glance. Reading values can be made more precise by using a ruler and drawing horizontal and vertical lines to the point on the graph whose values you want to find.

To plot a graph like this, two scales are needed. One is drawn along what is called the *X-axis* of the graph. It is the direction which is horizontal when the book you are reading is held upright. This X-axis has marks equally spaced along its length to represent equal steps of a quantity. In this particular case the quantity is current values.

The other axis is called the *Y-axis*. This is drawn vertically (at 90° to the X-axis), and in this example is used to plot values of voltage. Equal spaces along this scale thus represent equal steps of voltage. Both of these axes are illustrated here, left.

To plot the graph line itself, you choose a pair of corresponding values of *V* (volts) and *I* (amps), and locate the values of each on the appropriate axis. You then draw a line vertically upwards from the location of the selected current value on the X-axis. You then draw another line horizontally from the location of the selected voltage value on the Y-axis. Where these two lines meet, make a lightly pencilled dot to form the first point of the graph. The position of this dot represents the pair of values, one of current and one of voltage (see Figure 5.5).

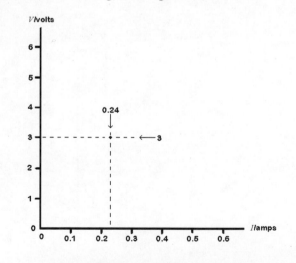

Figure 5.5   Plotting a point from a table of values

Other pairs of values are then similarly located and plotted to form a set of dots. These are then joined by a firmer line, drawn as smoothly as possible. This is a graph line. Every point on this line represents a pair of values, in this example, of voltage and current.

When the dots that you join in this way produce a *straight* line, the graph is said to be *linear*. We can say also that the relationship between the two values that being plotted against one another are linear. A linear graph is of great importance because:

It can be extended in either direction merely by using a ruler.

It means that one quantity is directly proportional to the other.

The example that was illustrated in Figure 5.4 is a linear graph that shows that the voltage across a resistor is proportional to the current flow through the same resistor. This relationship can also be expressed as an equation which we know as *Ohm's law*.

We usually label the axes of a graph with a symbol, oblique stroke and unit quantity. For example, the label *V*/Volts is read as 'V, in units of volts', and *I*/Amps as 'I, in units of amperes'.

The graph point 0,0 meaning the point at which both X = 0 and Y = 0 is called the *origin* of the graph. This is the point at which both of the quantities that we have plotted on the graph have zero values. The graph line in Figure 5.3 passes right through this point, meaning, in this example, that there is no voltage across a resistor when no current flows through it.

If, however, we plotted a graph of the voltages across a rechargeable cell (such as a nicad cell) against the current that was passing through the cell during the charging period (see Figure 5.5), the graph line would again be straight but it would not pass through the origin. This is because a voltage exists across the cell even when no charging current is flowing.

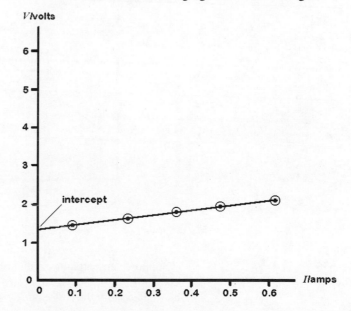

Figure 5.6　A graph with an intercept

This is an example of a graph that has an *intercept*. The intercept, in this example, is on the Y-axis, which means that there is a value of Y (voltage) when current $X = 0$. The intercept is this value of Y. In practical terms, this is the voltage of the cell when the charger is switched off.

Another quantity that can be read from a linear graph is its *slope*. The slope of a linear graph is found by taking two points, well spaced from one another, on the graph line, and finding their X and Y values. In Figure 5.5 the pairs of points that have been taken are indicated as X_1; Y_1, and X_2; Y_2, and their values are as shown on the figure. The lower value of Y_1, is then subtracted from the higher value Y_2, to give $Y_2 - Y_1$; and the X-values are subtracted in the same way, to give $X_2 - X_1$,. The ratio $\frac{Y_2 - Y_1}{X_2 - X_1}$ is the slope of the graph. In Figure 5.7 this slope has a value of 7.143, and we can round this to 7.1.

> The values of X and Y are always read from the *scales* marked along each axis, never from any measurement of distance made with a ruler. Only a linear graph has a single value of slope, a curved graph has a different slope value at each point on the graph.

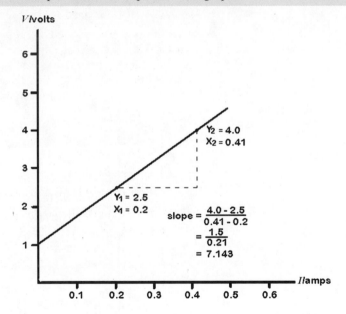

Figure 5.7   The slope of a graph

Suppose, now, that the value of the slope (m) and of the intercept (c) are known for any *linear* graph. The value of Y for any value of X can then be calculated from the formula:

$Y = mX + c$

Example: A graph has a slope value of 5 and an intercept of 2. What is the Y value corresponding to $X = 3$?

Solution: Substitute the data in the equation: $Y = mX + c$. Then:

$Y = (5 \times 3) + 2 = 15 + 2 = 17$

Note that in the working out of this example the multiplication was carried out before the addition. The general rule is that when you work out anything that consists of more than one action, all multiplications and divisions are carried out *before* any additions or subtractions. In olden days, this was memorized in the form MDAS (**M**y **D**ear **A**unt **S**ally), meaning multiplication, division, then addition, subtraction.

Proportion

We can use *proportion* on quantities that provide a linear graph when they are plotted. For example, the graph of voltage against current is a straight line, so that proportion can be used. If, for example, we know that a voltage of 10 V provides a current of 0.2 A, then a voltage of 20 V will cause a current of 0.4 A. In this example, both quantities have been multiplied by 2. Quantities like this are proportional, and we can write this in a shorthand way as $Y \propto X$, read as 'Y is proportional to X'.

Direct proportion like this can also be expressed in a formula:

$$\text{New } Y = \frac{\text{Old } Y \times \text{New } X}{\text{Old } X}$$

For example, if the old values of X and Y were 32 and 5 respectively, then the new X value is 47, then the new Y value is $(5 \times 47)/32 = 7.34$ (rounded).

Proportion can be used in this way only when the quantities would plot as a linear graph.

Question 5.2

If 0.1 A flows through a resistor when the voltage across the resistor is 2.5 V, what current will flow when the voltage is 4.0 V?

Waveform graphs

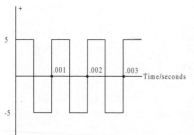

In this important type of graph, the quantity that is plotted along the X-axis is always *time*. The Y-axis is extended below the X-axis so that positive values can be plotted above the X-axis and negative values below it, as the Y-axis in the example, left, illustrates.

A waveform graph is used to show what values of voltage or current are present in a circuit at different times after the circuit has been switched on. The example shows a voltage which is alternately switched on and off, forming a type of waveform called a *square wave*.

Figure 5.8(a) also shows a square wave, but this time the voltage alternates between positive and negative values, with the time taken for a complete cycle of change (the period of the wave) from positive to negative and back to positive again only 0.001 seconds.

Figure 5.8(b) illustrates a very common type of waveform called a sine wave, which changes smoothly from positive to negative values and back again to positive, in a time (for the example shown in this illustration) of 0.02 seconds. This time is also called the *period* of the wave. This particular type of sine wave is typical of the mains (line) supply voltage used in the UK.

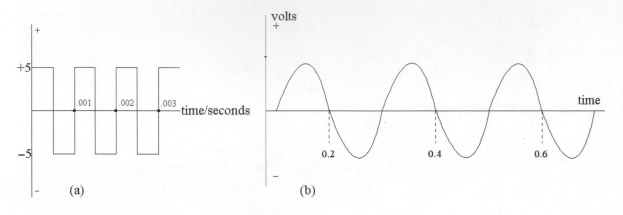

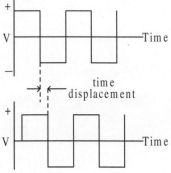

Figure 5.8   (a) A square wave with both positive and negative portions, (b) a typical sine wave

The drawing, left, shows a pair of square waves taken from different parts of a circuit, but plotted together on the same time scale. It will be seen that the waves, though taking the same time to complete one cycle, are not in step with each another. Such waves are said to have a time or *phase* displacement relative to each other.

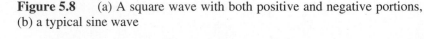

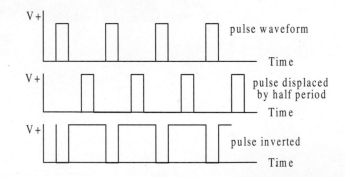

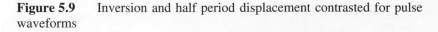

Figure 5.9   Inversion and half period displacement contrasted for pulse waveforms

Figure 5.9 shows an example of waves which are of the same general shape, but which are nevertheless not identical. In the case shown, one wave is inverted, upside down, compared to the other. With waves like sine waves and square waves, there is no difference in shape between waves which are shifted by a half period and those which are inverted. For other waveshapes, the differences are both marked and important, as can be seen in Figure 5.9.

Frequency and frequency response

A wave has a shape that repeats, and the frequency of a wave is the number of times the shape repeats in a second. This is defined as the number of complete cycles of the wave from its positive peak to its negative peak and back again (or the reverse), that occur per second. This frequency can be calculated from a waveform graph by reading off the time period (T) of one cycle and then calculating $1/T$ (the reciprocal of the value of T), which is the frequency. Figure 5.10.

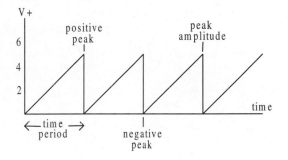

Figure 5.10 Time period and peak amplitude

Instruments such as the oscilloscope will display a waveform graph on the screen, and you need to be able to find the frequency of the wave from this display.

> The negative peak of a wave might not be an actual negative voltage, only a voltage which is lower than the voltage of the positive peak. For example, the positive peak might be at +6 V and the negative peak at +2 V.
>
> *Example*: The time period of a wave is 0.05 s. What is the frequency of the wave?
>
> *Solution*: $1/T$ in this case is $1/0.05$, or 20 complete cycles of wave per second. The unit of frequency is the hertz (shortened to Hz), so the frequency in the example above would be written as 20 Hz.

The *peak amplitude* of a waveform is the maximum value of the quantity that is plotted on the Y-axis be it voltage or current, positive or negative. For example, the amplitude of the voltage in the waveform of Figure 5.8(a) is 5 V, while the amplitude of the waveform plotted in Figure 5.10 is also 5 V.

In a *frequency response* graph, amplitude is plotted on the Y-axis and frequency on the X-axis, so that the graph can be used to read off the amplitude of a waveform at a number of different frequencies. When such a graph is plotted, it is usual to assume that a circuit such as an amplifier is fed with waveforms having the same amplitude, whatever their frequency. The graph then shows the amplitude of the wave at the output of the circuit at every frequency which has been applied. An example of the shape of such a graph is shown later in Figure 5.11.

A frequency response graph of this kind is an important factor in deciding whether an amplifier is suitable for its planned application.

Non-linear graphs

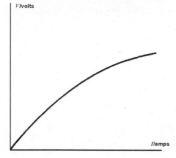

Not all graphs are linear. For example, a graph of current plotted against voltage in a component called a *thermistor* would take the shape shown in the drawing, left.

Graphs that are curved like this have no single value of slope – each point on the graph line has a different value of slope. These graphs are useful for finding values of I and V which lie within the range of the graph line, but the graph line itself cannot be extended except by guesswork. In addition, you cannot easily calculate values from a formula.

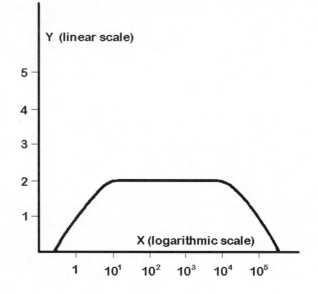

Figure 5.11   A graph with a single logarithmic scale

Sometimes a straight-line graph is obtained by the use of scales which are themselves not linear. A non-linear scale is one in which equal distances along the scale do not represent equal amounts of the quantity being plotted.

The graph in Figure 5.11, for instance, uses a logarithmic scale on which each equal step of distance represents a tenfold increase of the quantity being plotted. Logarithmic scales are often used for plotting the frequency of electronic signals, because a very large range of values can then be plotted on a single graph.

One disadvantage of a logarithmic graph is that it is never easy to locate intermediate values that lie between the printed values on the axes. For example, the point that represents a value of 50 along the X-axis in Figure 5.11 is *not* midway between 10 and 100 nor is it midway between 1 and 100. Another disadvantage is that the value of the slope of a logarithmic graph is of little use even if the graph happens to be a straight line

Inverse and exponential law graphs

The inverse law shape of graph is found when one quantity is related to the inverse of another. For example, the inverse of X is $1/X$, so that a graph of $Y = 1/X$ would take the shape illustrated in Figure 5.12(a).

Another important graph shape is the exponential of Figure 5.12(b). In this type of graph, each step of value added to the one axis results in a large change in the value on the other axis. In the illustration, each small change in current requires a progressively larger step in time.

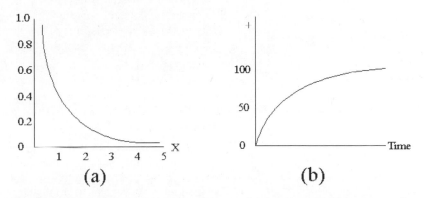

Figure 5.12   (a) An inverse law graph, (b) an exponential graph shape

Answers to questions

5.1   5 V r.m.s.

5.2   0.16 A.

6 Capacitance, capacitors, inductance and inductors

Capacitance between plates

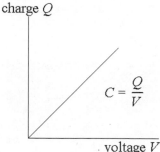

charge Q

$$C = \frac{Q}{V}$$

voltage V

A simple capacitor can be made by using two conducting plates separated from each another by a (non-conducting) gap. When the plates are connected to a battery a current will flow, dying down to zero in a very short time. The action of the battery has been to pump electric charge (electrons) from one of the plates on to the other until the latter is so negatively charged that it will take no more. It is this pumping action, a movement of electrons caused by the potential of the battery, that is detectable as momentary current flow. Because one plate is now negatively charged the other is positively charged, because electrical charges always exist in pairs.

If the capacitor is now disconnected from the battery, the electrons that have been moved from one of its plates to the other will still remain out of place, but will return to the more positive plate through any conducting path that connects the plates. This return movement will cause another momentary current to flow, this time in the opposite direction.

While it is connected to the battery and from a short time after it has been disconnected from the battery the capacitor, is said to be *charged*. The amount of charge that is stored depends on the voltage of the battery (or other supply) that is used to do the charging, and to another quantity called *capacitance* which is a measure of the ability of the capacitor to store electric charge.

While a capacitor is being charged or discharged, the ratio of the charge transferred *between* the plates to the voltage *across* the plates remains constant, so that a graph of charge plotted against voltage is a straight line. In symbols, $C = Q/V$, where Q is the charge measured in coulombs and V the voltage in volts. The constant C is the capacitance which is measured in *farads* (F) when the other units are quoted in terms of coulombs and volts.

The coulomb is a very large unit. It is the amount of charge carried by 6.28×10^{18} electrons and so the farad is also a very large unit. For this reason, the subdivisions of the farad that we call the microfarad (μF), the nanofarad (nF) and the picofarad (pF) are used in electronic work. One μF $= 10^{-6}$ F (one millionth of a farad), and 1 nF $= 10^{-3}$ μF (one thousandth of a microfarad). The pF is one thousandth of the nanofarad, and thus one millionth of the μF and 10^{-12} F.

The value of the capacitance which can be achieved by the arrangement of two parallel conductors depends on three factors:

- the overlapping area of the conductors,
- the distance they are apart, and

- the type of insulating material (or *dielectric*) that is used to separate them.

These factors affect the capacitance value in the following ways:

Capacitance is proportional to the area of the conductors that overlap – this is their effective area for purposes of capacitance (see Figure 6.1). The greater the area of overlap, the greater the capacitance. Some types of capacitor form comparatively large values of capacitance by using conductors in the form of strips which can be wound into rolls, so that a large amount of capacitive area is obtained in a small space.

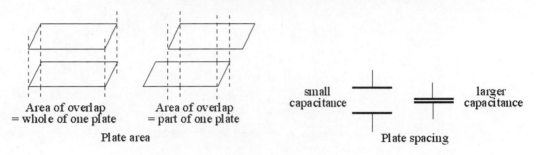

Area of overlap = whole of one plate   Area of overlap = part of one plate

Plate area

small capacitance   larger capacitance

Plate spacing

Figure 6.1   Capacitor plate overlap and spacing

Capacitance is inversely proportional to the spacing between the conducting plates. This means that, for a given pair of plates, a reduction in the distance between them *increases* their value of capacitance. High value capacitors therefore need insulators that are as thin as possible. Thin insulators, however, are easily broken down if the voltage across them is too high. Every capacitor therefore carries a maximum *voltage rating*, which must not be exceeded. Failure of the insulating material in a capacitor is called *dielectric breakdown*. It results in the failure of the capacitor, often producing a short circuit between the terminals.

The type of material that is used as an insulator in a capacitor also contributes to its value of capacitance. When air or a vacuum is used as an insulator, capacitance between a given pair of plates is at its lowest possible value. Using materials such as waxed paper, plastics, mica and ceramics can increase the capacitance up to 25 times the value given by the same thickness of air or vacuum. This greater efficiency is caused by an effect that we measure as the *relative permittivity* of the material.

$$C = \frac{\varepsilon \times \text{area}}{\text{spacing}}$$

or

$$C = \frac{\varepsilon A}{d}$$

The formula, left, summarizes the three factors governing a capacitor's maximum value of capacitance. Units are capacitance in farads, area in square metres and spacing in metres. The factor ε (epsilon) is the permittivity. This can be put into more practical units, so that for plates separated by air the formula becomes as shown left with A in square metres, spacing d in mm, and ε equal to 0.0088.

Capacitor types

Many materials and methods of construction have been used to make capacitors, but only a few representative types can be usefully considered here.

Variable capacitors operate by changing the area of overlap between the two plates of a capacitor, or by altering the spacing between the capacitor plates. The old-fashioned multiple plate variable capacitor (seldom seen now) has one set of blades fixed and an interleaving set mounted on a rotating shaft, see Figure 6.2. As the plates are more fully meshed, so the area of overlap becomes greater and the value of capacitance is increased (and vice versa).

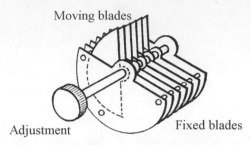

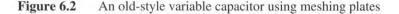

Figure 6.2   An old-style variable capacitor using meshing plates

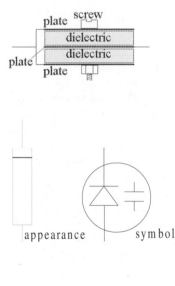

When the dielectric is air, a large capacitor is needed even for a capacitance of only 500 pF, so for making miniature capacitors the plates are separated by sheets of solid dielectric. In *compression trimmers*, illustrated left, most of which have a capacitance of some 50 pF maximum, sprung plates are separated by solid dielectric sheets. An increase in capacitance is obtained by compressing the plates towards each other by means of a screw. Variable capacitors such as these are designed with the moving (adjustable) plates always connected to earth, so as to avoid the changes of capacitance which would otherwise occur when the control was touched; an effect called hand or body capacitance.

Another type of variable capacitor is the *varicap diode* whose appearance and symbol are illustrated here. This device is a semiconductor diode whose capacitance is varied by changing the voltage across it. This is used now to allow capacitance changes by remote-control. See also Chapter 8.

Activity 6.1

Use a capacitance meter set so as to measure values of up to 1 000 pF, and measure the capacitance between two metal sheets separated by insulators of different thickness. Use sheets of different area to show that the capacitance value depends on area. Try to keep the arrangement of wiring unchanged to avoid effects of stray capacitance.

Small capacitors of fixed value (see left) are simply constructed, with the plate and the insulator arranged parallel to one another. Their insulators are often thin sheets of mica, which can be silvered on both sides to form a silver mica capacitor. This is a type that is chosen when stability of

capacitance value is important. Thin sheets of ceramic are also used. The complete capacitor, with its leadout wires attached, is then dipped in wax or plastic so as to insulate the plates and protect the assembly from moisture.

An alternative construction is the tubular type, in which the inside and outside surfaces of a ceramic tube are coated with metal to form a capacitor (Figure 6.3) This type is also coated in wax or plastic.

Question 6.1

Why does a capacitor need wax or other sealing materials to prevent moisture penetrating?

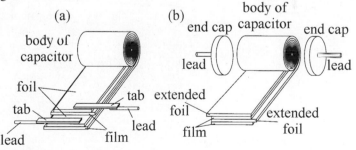

Figure 6.3   A tubular ceramic capacitor (enlarged view)

In wound or rolled capacitors, the conductors take the form of foil sheets. Typically, these are formed by evaporating metal on to both sides of a long strip of insulator, placing an insulating sheet along one of the sides to prevent short circuits, and then rolling the whole assembly up like a Swiss roll. Separate metal foils can also be rolled together in this way, as shown in Figure 6.4.

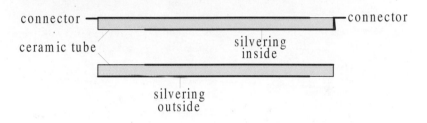
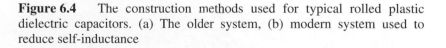

Figure 6.4   The construction methods used for typical rolled plastic dielectric capacitors. (a) The older system, (b) modern system used to reduce self-inductance

Nevertheless, rolled capacitors using insulators that can withstand high voltages (high dielectric breakdown values) are used for manufacturing capacitors that must be of fairly high capacitance values and also withstand high voltages.

The insulators are either waxed paper or plastics of various kinds, such as polycarbonate, polythene or polyester. This type of construction can be used for values of 1 nF to 1 µF or more, but for larger values of capacitance rolled capacitors are bulky.

Electrolytic capacitors

Electrolytic capacitors are constructed differently using thin films of aluminium oxide as the insulator. A typical electrolytic capacitor is made using sheets of aluminium foil that are separated by a porous material soaked in an acid solution (the electrolyte). When a voltage is first applied across the plates a current flows briefly until one plate is covered with an invisibly thin film of aluminium oxide. This process is called *forming*. The oxide film is a good insulator, and since it is thinner than any other useable solid material, electrolytic capacitors can have very large values of capacitance. In addition, the area of the foil can be greatly increased by corrugating it or, better still, by etching it to a rough finish. The voltage which can be applied across the film is, however, limited; and it must be applied in the correct polarity.

One of the terminals of an electrolytic capacitor is marked (+), and the other (−); and this polarity *must* be observed. Connecting an electrolytic capacitor the wrong way round would cause the oxide film to be broken down, and large currents would flow. This could cause the casing to burst, spraying both the operator and the rest of the circuit with corrosive material. Electrolytic capacitors used to smooth power supplies are particularly at risk in this respect, for very large currents would flow in the event of an internal short-circuit.

The illustration shows two capacitors with radial leads and one with axial connections. Generally the radial lead types are used with the larger components whilst the axial leads need to be bent and cropped before insertion into a PCB.

Electrolytic capacitors are used for power supplies, see Chapter 9. They are also used as coupling and decoupling capacitors. In both of these applications, the capacitor allows a.c. (signal) currents to pass freely, even if the signal frequency is low, while preventing the passage of all but a very small d.c. *leakage* current. If any d.c. leakage current is undesirable, then capacitors of a different construction must be used. Electrolytic capacitors should not be connected in series, because the leakage currents are likely to be different, and this might cause the voltage across one capacitor to be much larger than the voltage across the other. In the few examples where this has to be done, the capacitors will have resistors connected in parallel.

Question 6.2

Smoothing capacitors on some radio transmitters use paper or ceramic construction. Why are electrolytics not used?

Tantalytics

For some purposes, a useful alternative is the *tantalum electrolytic* (or *tantalytic*), which permits very much less leakage current but is a more expensive component. Tantalum electrolytics, unlike the aluminium variety, can be obtained in either polarized or unpolarized form.

Reliability of capacitors

Capacitors nowadays are regarded as the least reliable components in electronic circuits, and they are the first items to be checked when a problem arises. Electrolytic capacitors are particularly likely to cause trouble, and many servicing procedures specify that all electrolytic capacitors are replaced on servicing just as a matter of course. Failure of the power supply smoothing capacitor is a common source of microwave oven breakdowns.

Wound capacitors with values in the range 1 nF to 1 F are used in audio coupling, decoupling and filtering, in RF. decoupling and in low frequency oscillators. Ceramic capacitors are used in RF decoupling, and silver mica capacitors in RF tuned circuits. Variable capacitors are used in oscillator, tuned amplifier and tuned filter circuits.

Stray capacitance

In addition to any capacitors which may be deliberately placed in a circuit, a *stray* capacitance exists between any two parts of a circuit that are physically close to one another but not connected. Stray in this sense means unintentional and unplanned.

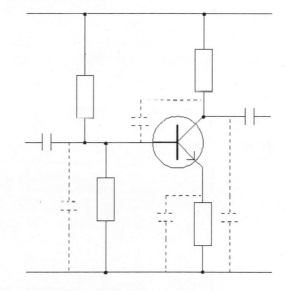

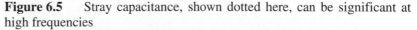

Figure 6.5 Stray capacitance, shown dotted here, can be significant at high frequencies

When the circuit is used with low-frequency signals, this stray capacitance is not very important; but things can be very different in circuits working at high frequencies. If, for instance, the circuit shown in Figure 6.5 were required to handle high frequency signals, the stray capacitances that are indicated by dotted lines could cause unwanted effects.

Stray capacitances can often be minimized by careful layout of a circuit, but there is *always* an unavoidable stray capacitance across each individual component and this amount is not reduced by altering the layout.

Activity 6.2

Use a capacity meter to show the effects of stray capacitance. Observe the changes in capacitance that appear when the meter is connected between a metal plate and earth, and you place your hand near the plate.

Screening

Screening makes indirect use of the effects of stray capacitance to shield one part of a circuit from signals radiated by another part. An electrostatic screen removes unwanted signals which could be passed by stray capacitances by enclosing the circuit which could be affected inside an earthed metal box. This box need not be solid, but can be made using metal mesh. Any stray signal current reaching the box returns to earth through the metal casing, and is so prevented from affecting the circuits being screened.

A different form of screening is needed for magnetic signals. It is described later in this chapter.

Capacitors in combination

Capacitors, like resistors, can be connected in series, in parallel, or in series parallel combinations. Whatever the method of connection, any combination of capacitors can be replaced, as far as calculations are concerned, by a single capacitor of equivalent value. The size of this equivalent capacitance can be calculated from first principles. It is equal to the total charge stored by the capacitor network, divided by the total voltage across the network.

This method of calculation must be used in complex capacitor circuits, but simple formulae can be used to calculate the resultant capacitance when capacitors are connected either wholly in series or wholly in parallel (see Figure 6.6).

When capacitors are connected in parallel, their equivalent value is found by adding their individual values of capacitance. Thus

$$C_{total} = C_1 + C_2 + C_3 + \cdots$$

When capacitors are connected in series, their equivalent value is found by adding the inverses of their individual values of capacitance. Thus

$$\frac{1}{C_{total}} = \frac{1}{C_1} + \frac{1}{C_2} + \frac{1}{C_3} + \cdots$$

or for two capacitors

$$C_{total} = \frac{C_1 C_2}{C_1 + C_2} \quad \text{(product divided by sum)}$$

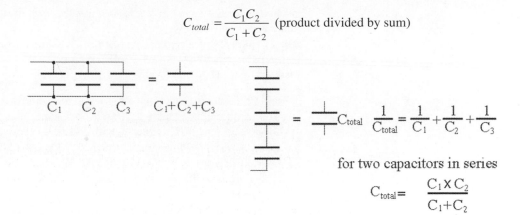

for two capacitors in series

$$C_{total} = \frac{C_1 \times C_2}{C_1 + C_2}$$

Figure 6.6   Capacitors in parallel and in series

These formulae for capacitors in combination are at first sight similar to those governing resistors in combination. The important difference is that the formula for adding resistors *in series* is the same as that for adding capacitors *in parallel*, while the formula for adding resistors *in parallel* is the same as that for adding capacitors *in series*.

Time constant

The voltage across the plates of a capacitor rises as charge flows into the capacitor from a source such as a battery. When the voltage across the plates of the capacitor is equal to the battery voltage, charge cannot continue to flow because there is no voltage difference to cause electrons to move. The capacitor is now completely charged.

When the charging circuit contains a resistor in series, the rate of movement of charge (electron flow) is reduced. Rate of movement of charge is just another name for current flow, and you would expect the presence of a resistor in series in the circuit to reduce current flow. Adding more resistance to a capacitor charging circuit therefore increases the time that is needed to charge the capacitor.

In the same way, if the capacitor is disconnected from the power supply and a resistor is connected in parallel with it, charge will return through the resistor until the capacitor is discharged again, with no voltage difference remaining between the plates. The resistor will, however, slow down the rate of discharge of the capacitor.

In practice, this charge and discharge process is normally so fast that a CRO is needed to observe the changing voltages across the capacitor plates.

A graph of voltage plotted against time for a charging capacitor is shown, left. This graph is an important one. Its shape is the type called *exponential* and the notable feature about it is the way in which the graph flattens out as the plates of the capacitor approach their final value of voltage difference. Because of this flattening out effect, the final value of voltage is only

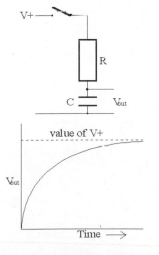

gradually approached, with the last percentage of the charge taking a relatively long time to get on to the plate. In other words, the capacitor is almost charged for a relatively long time before it is fully charged and the last small percentage of its full charge is relatively unimportant in measuring its effective capacity.

For this reason, a quantity called the *time constant* of the capacitor–resistor combination is used as a measure of the length of time which the capacitor needs to be charged to within this last few per cent of its final value of charge. The time constant is calculated by multiplying the R and C values, so that as an equation: $T = R \times C$. When R is quoted in ohms and C in farads, the quantity RC will be given in units of seconds of time. When, as is more usual, resistance is measured in kilohms (K) and capacitance in nF, the time constant is given in microseconds (μs).

The importance of the time constant is that charging or discharging is about 63 per cent complete after the elapse of one time constant. After a period of three time constants, charging is 95 per cent complete, and we can take this time as the time for complete charging or discharging. For a few applications, charging or discharging is taken to be complete after four time constants (98 per cent charged).

The drawing, left, shows the effect of *discharging* a capacitor. You can see that the graph shape is still exponential (though the other way round). The same comments about the practical value of a time constant therefore apply to capacitor discharge as they did to charge, with the charge reduced to 37 per cent of its initial value after one time constant, and to about 5 per cent of its initial value after three time constants (or to 2 per cent in four time constants).

Figure 6.7 shows universal charge/discharge curves which enable the voltage of a charging or discharging capacitor to be read off the chart in terms of the capacitor's time constant and maximum voltage.

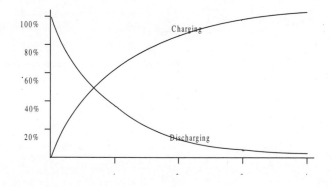

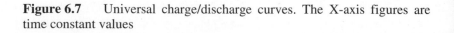

Figure 6.7   Universal charge/discharge curves. The X-axis figures are time constant values

The importance of these charging and discharging time constants is, as you will see later, that they can be used in two forms of wave shaping operations, known as differentiation and integration, which can be performed by resistor capacitor networks of the type described.

Question 6.3

What is the time constant for a 4K7 resistor and a 100 nF capacitor connected together?

At signal frequencies, capacitors permit signal current to flow when a signal voltage exists between the plates. This current is the charging and discharging current of the capacitor. The ratio of signal voltage to signal current is called the *reactance* (X_c) of the capacitor, and is measured in units of ohms. Figure 6.8(a) shows a graph of capacitor reactance plotted against capacitance, for a constant frequency of signal; while Figure 6.8(b) shows a graph of capacitor reactance plotted against frequency, for a constant value of capacitance.

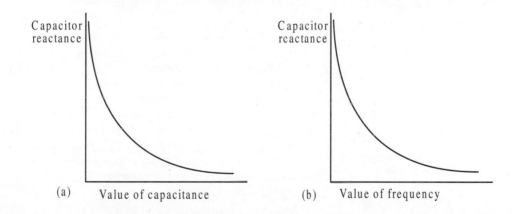

(a)   Value of capacitance   (b)   Value of frequency

Figure 6.8   (a) Capacitor reactance plotted against capacitor value for a fixed frequency. (b) Capacitor reactance plotted against frequency for a fixed capacitor value

The graphs show that the reactance of a capacitor is lower for large values of capacitance than it is for small values, and is also lower at high frequencies than it is at low frequencies.

The reactance of a capacitor can be calculated from the formula:

$$X_C = \frac{1}{2\pi fC}$$

and the units of reactance are ohms if C is in farads and frequency is in Hz. For units of nanofarads and kilohertz, the reactance will be in units of megohms. For units of nanofarads and MHz, the reactance units are kilohms, and this combination is particularly useful for calculations.

A perfect capacitor would have zero self inductance, together with zero series resistance and shunt leakage so that the phase angle between the applied voltage and the resulting current would be 90° with *I* leading *V*.

Any self-inductance due to the nature of the construction would create a small phase shift that would counteract the true reactance. Similarly any resistance would dissipate energy and contribute to heat losses.

In the sketch, left, the equivalent shunt resistance losses which would in any case normally be very small, have been neglected and the series resistance which in general consists of the total resistive effect is known as the effective series resistance (ESR).

The phasor/vector representation shows the relationship between these reactive effects with the self inductive component cancelling some of the capacitive reactance which affects the measured value of the device. An increasing series resistance causes the phase angle θ to fall below the theoretical 90° value and the so called loss angle δ (or tan δ) to rise. This increases the impedance of the capacitor in spite of its value remaining virtually constant.

Alternating current flowing through this practical capacitor dissipates heat energy in the ESR component and this in turn, increases the leakage current. This roughly doubles for every 10°C rise in temperature. In the working environment the ambient temperature will also rise and together this can lead to premature failure. Unfortunately, a failed capacitor usually leads to the failure of other components as well. Hand held test meters are now available that will measure both the value of capacitance and the ESR which typically could fall between 0.1Ω and 1Ω depending on the type of capacitor and its value. However, care needs to be exercised because a capacitor whose value falls within tolerance can easily exceed an acceptable value for ESR.

The drawing also shows how the self-inductance of the device can combine with the capacitor to produce a self resonant frequency to form a further limiting feature. Capacitors are also temperature limited and typically have working values that range from 45°C to 125°C. Replacing a failed capacitor rated at say 85°C with one rated at 105°C will usually improve the reliability of a system.

The practical capacitor

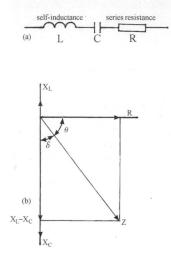

(a)   L   C   R

self-inductance   series resistance

(b)

Charge retentivity

A capacitor that has been charged and then discharged particularly by a short circuit will after a while, recover some of its lost charge. This is often described as voltage remanence, retentivity or soakage and is chiefly due to the distortion of the atomic structure of the dielectric in the charged state. After apparently being discharged, the atomic structure relatively slowly returns to its resting state to produce the retentive effect.

Question 6.4

What is the reactance of a 47 nF capacitor at 10 MHz?

Magnetic flux

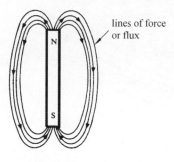

lines of force or flux

When iron filings are sprinkled on to a sheet of paper that has been placed over a bar magnet, the filings will arrange themselves into a definite pattern, which we call a *flux pattern*. This flux pattern, illustrated left, is a map of the direction and strength of the forces which are exerted on the iron filings because they are close to the magnet. The lines that the filings form indicate the direction of the force. The clumping together of filings is some indication of the strength of the force.

These forces exist and can be detected without any physical contact between the magnet and the iron filings. The general name for a force of such a type is a *field* force. The flux pattern, therefore, is a map of the *magnetic field* of the magnet, and it indicates the direction and strength of this field.

The pattern of the field which exists around a magnet can be changed in several ways. Figure 6.9 shows how the flux pattern appears when two magnets are placed close to one another. When both magnets are aligned in the same North–South direction, the pattern is as shown in diagram (a). When one of the magnets is reversed, the pattern alters to that shown in diagram (b).

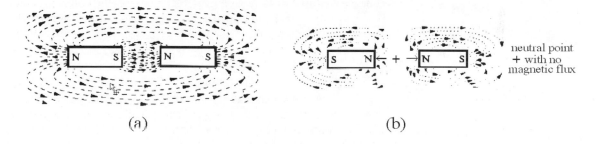

neutral point + with no magnetic flux

(a)   (b)

Figure 6.9   The flux pattern around two adjoining magnets (a) N to S and (b) N to N

The shape of flux patterns is governed by two rules:

A line behaves as if it were carrying a magnetic 'current' (a *flux*) in the direction of the line. This direction can also be indicated by a compass needle.

The flux lines become closely spaced together when they are close to magnetic material so indicating a strong field, but they are more widely spaced elsewhere, where the field is weaker.

The shape of a flux pattern is not noticeably altered by being close to non-magnetic materials such as plastics, copper or aluminium. Placing magnetic materials in a magnetic field, however, causes the flux pattern to change as shown in Figure 6.10. The magnetic material behaves as an easy path for the lines of flux, just as a metal offers an easy path for current. This allows us to manipulate and control lines of flux in what is called a magnetic circuit, a topic that is important for all magnetic devices such as motors, relays and solenoids.

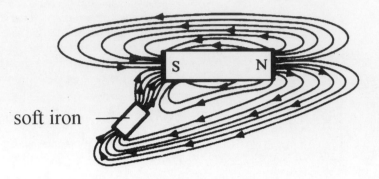

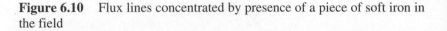

soft iron

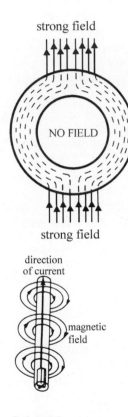

Figure 6.10   Flux lines concentrated by presence of a piece of soft iron in the field

Some metal materials that we call *soft* magnetic materials can concentrate flux lines although they are not permanent magnets themselves. These metals (soft iron, Permalloy and Mumetal are examples) can be used as *magnetic screens*. The illustration, left, shows that a cylinder of soft magnetic material will have no flux pattern inside the cylinder even when there is a strong field outside it. Such screening materials are used to shield cathode ray tubes and other components from stray magnetic fields.

A *soft* magnetic material is one that is easily magnetized by another magnet, but which loses that magnetism equally easily. A *hard* magnetic material (such as steel and several steel alloys with cobalt and nickel) is difficult to magnetize, but it will retain its magnetism for long periods unless it is heated, struck with a hammer, or de-magnetized by alternating fields. Hard magnetic materials are used for making permanent magnets.

Permanent magnets are not the only source of magnetic flux. When an electric current is passed through any conductor, a magnetic flux pattern is created around the conductor. The shape of this flux pattern around a straight conductor, illustrated, is circular, unless it is distorted by the presence of other magnetic material, and the pattern has no start or finish points. This leads us to the conclusion that *all* flux lines are, in fact, closed lines without either a start or a finish.

A conductor that is wound into the shape of a coil is called a *solenoid*. When current flows through the wire of a solenoid, the flux patterns of each part of the wire in the solenoid add to one another, so producing the pattern shown in Figure 6.11. This pattern is similar to that of a bar magnet except that the shape of the flux pattern *inside* the coil can be seen.

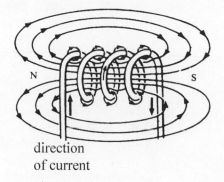

direction
of current

Figure 6.11   The flux pattern around a solenoid carrying current

Activity 6.3

Use a piece of white cardboard and some iron filings to show that the field pattern caused by a bar magnet is the same as that of a solenoid carrying current.

Question 6.5

Given a solenoid and a permanent bar magnet of equal strengths, what is it that makes the solenoid more useful for some applications but less useful for others?

The addition of a soft magnetic material as a core inside a coil greatly concentrates the flux pattern at its ends and therefore inside the core also. The concentration of the flux pattern means that a coil becomes a much stronger magnet when it has a soft magnetic core though only for as long as current is flowing through the wire of the coil.

A suitable coil material can in practice increase the strength of a magnetic field several thousand-fold. This allows us to make electromagnets which are very much stronger than any permanent magnet could possibly be.

All magnetic flux is caused by the movement of electrons. This can be the movement of electrons through a conductor which we call electric current, or it can be the spinning movement of electrons within the atoms which occurs in permanent magnetic materials. These spinning movements balance each other out in most materials (so that most materials are only very faintly magnetic), but the materials which we class as strongly magnetic (called *ferromagnetic* materials) are made up of atoms which have more electrons spinning in one direction than in the opposite direction.

Another important type of alteration in the flux pattern of a magnet occurs when a wire is placed in a magnetic field and has a current flowing through it. In Figure 6.12, N and S are the North and South poles, respectively, of two bar magnets. The little circle with a cross inside it which lies between them represents a conductor wire seen end on, that is to say, with one of its ends running into the paper away from you, and the other end running out of the paper towards your eye.

When an electric current is passed through the wires the flux pattern between the magnets becomes distorted and this pattern can be revealed using iron filings or small compass needles.

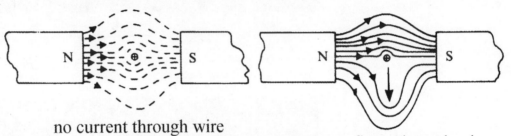

no current through wire

current flows through wire
(away from you into paper)

Figure 6.12   What happens to flux lines when current flows through a wire in a magnetic field

Left-hand rule

When current flows and causes this distortion of the field, a magnetic force is exerted on the wire itself, acting at right angles to the wire and also at right angles to the flux lines. The direction in which this force is exerted can be remembered by using the left-hand rule as pictured in Figure 6.13. The thumb and first two fingers of the left hand are extended at right angles to one another. The first finger then points in the direction of the flux (N to S direction); the second finger shows the direction of current flow (positive to negative); while the thumb points along the direction of the magnetic force exerted on the wire. Unless the wire is held firmly, it will move in the direction of this force. A familiar illustration of this effect is the way that the cables to an electric welder jump apart when the arc is struck – each cable generates a magnetic field and so exerts a force on the other cable.

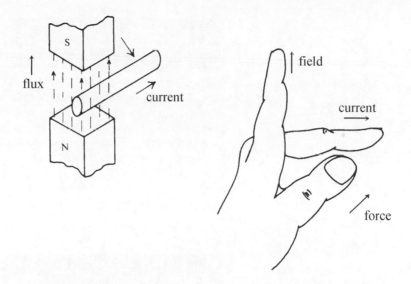

Figure 6.13   The left-hand rule for a conductor

Electric motor principle

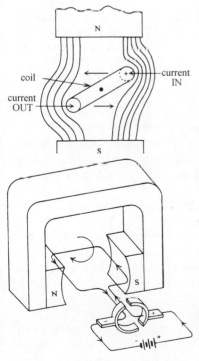

The force that exists between a magnetic flux pattern and a wire carrying current is the principle of the electric motor. When current flows through the coil of wire shown in cross-section, left, the ends of the coil will experience forces which are equal in size but opposite in direction. The coil will therefore rotate about its central axis. This rotation will only continue, however, until the plane of the coil lies in line with the flux (in what is called its neutral position). In this position no force is acting to cause rotation because the coil is now aligned with its flux pattern. Another way of saying this is that the coil axis no longer *cuts across* the flux lines.

Continuous rotation is possible if the direction of current flow through the coil of wire is reversed just as the neutral position is reached. The momentum of the rotating coil will carry it a little way past the neutral position, and the reversed current will exert a force on both ends of the coil which keeps the coil rotating.

The direction of current flow through the wire can be reversed at the correct point in its rotation by using a rotary switch called a *commutator*, see drawing, left. The segments of the commutator are made of copper, and the contact (or *brush*) is soft carbon, conducting current but allowing the movement of the commutator segments with fairly low friction.

An electric motor with a single coil and a two-segment commutator does not run at a perfectly uniform speed. The coil moves fastest when it is at 90° to the plane of the flux and slowest as it passes the neutral position. In practice, d.c. motors use several coils and corresponding pairs of commutator segments. This provides smoother running and avoids the problem of 'sticking' when the motor will not start if the coil happens to be at rest near its neutral position. Small motors usually have a twin coil arrangement with four commutator segments.

Activity 6.4

Use a simple two-pole d.c. motor and a strobe light to show that the rotation speed is not constant. Examine a multi-pole motor also.

The set of revolving coils is known as the *armature* of the motor, and each end of a coil is called a *pole*, so that a small motor can be described as having a 4-pole armature. The magnet (which may be either a permanent magnet or an electromagnet) is known as the *field* winding. D.c. motors are easy to control and their direction of rotation can be reversed by reversing the direction of current in either the armature or the field, but not both together.

Activity 6.5

Show using a simple d.c. motor with a wound field that the speed of the motor is controlled by the amount of field current. Does increasing the field current make the motor run faster or slower?

The moving-coil meter

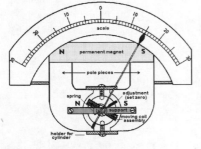

The moving coil meter uses the same principle as the electric motor to measure the d.c. current flowing through the coil. In this case, the coil is wound on a former with a pointer attached and mounted in bearings. The coil is positioned between the shaped poles of a powerful electromagnet. Current is fed to the coil through a pair of spiral springs which act to cause the coil to take up a zero position when no current is flowing.

When a current flows through the coil, the magnetic field of the coil works with the flux of the permanent magnet to turn the coil. The amount of turning is proportional to the amount of current flowing in the coil because the pointer stops at a position where the magnetic force is exactly balanced by the mechanical force of the springs.

Electromagnetic induction

The reverse of motor action is called *electromagnetic induction*. When the flux pattern around a conductor (with no current passing) is changed by *moving* the wire through the flux, a voltage is generated between the ends of the wire. This *induced voltage* is greatest when the direction of the flux and the movement of the wire are at right angles to one another. This is the basic electrical generator principle.

The polarity of the induced voltage can be remembered by using the *right-hand rule* illustrated in Figure 6.14 The first two fingers and the thumb of the right hand are extended at right angles to one another. The first finger is pointed in the direction of the field of the magnet (N to S), and the thumb in the direction of the motion of the wire. The second finger then indicates the end of the wire which is positive. If the ends of a wire are connected to a load, current will flow. Remember that the **F**irst finger points in the **F**ield direction, the **M**otion is in the thum**B** direction, then the se**c**ond finger **P**oints to **P**ositive (also the direction of current(I).

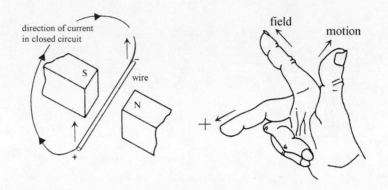

Figure 6.14 The right-hand rule for finding the polarity of an induced voltage

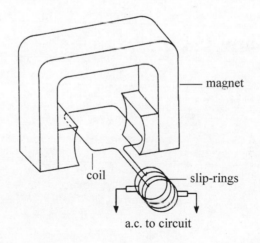

Figure 6.15 The principle of the a.c. generator, using slip-rings

Alternator

This principle of *motional induction* is used to generate electricity in an *alternator*. Figure 6.15 shows a simple alternator using the same construction as an electric motor but with the commutator replaced by slip-rings. These slip-rings do *not* reverse the connections to the coil. The waveform which is produced when the alternator is set spinning is a sine-wave whose voltage reaches maximum when the wire of the coil is momentarily moving at right angles to the flux, and zero at the instant when the wire of the coil is moving in line with (i.e. parallel to) the flux

+**volts**

time

**Transformer
induction**

direction. The reversal of polarity is caused by the fact that the wire during its rotation must cut across the flux lines in one direction for half of its rotation and in the opposite direction for the other half.

The voltage produced by an alternator increases as the speed of rotation increases. Large voltage outputs can be obtained by using large magnetic fields and many turns of wire in the rotating coil. The *frequency* of the a.c. output is determined *entirely* by the rotating speed of the spinning coil and the number of poles of the magnet.

If the slip-rings are replaced by a commutator, the waveform generated will be of the unidirectional form, illustrated. This is close enough to d.c. to be useful for some purposes, such as battery charging. Alternators that are fitted with several coils and slip-rings generate *multiphase* a.c., with each output wire carrying a sine wave which is out of phase with all the others. Remember that our electricity mains supply is three-phase, with house supplies obtained by connecting between one phase and neutral. Alternators for cars use a three-pole arrangement with a set of six diodes rectifying the output to a reasonably smooth d.c.

A d.c. generator that uses several coils and commutator segments, however, produces a reasonably smooth d.c. output, though the switching at the commutator segments produces sparks, causing spikes of voltage to appear. This causes interference, called *hash*, on car radios, and the use of alternators and diodes on car electrical systems was a major step forward in improving car radio reception.

It is also possible to generate a voltage by induction without any *mechanical* movement. If the strength of the flux that cuts across a conductor is varied, a voltage will be induced in the conductor, just as if the variation of flux had been caused by movement. This effect, called *transformer induction*, is responsible for the very important electrical effects known as *mutual-inductance* and *self-inductance*.

When a current starts to flow through a coil of wire, or is switched off, the flux lines around the coil will respectively expand or collapse. This variation in the flux lines induces a voltage in the coil itself and this induced voltage is always in a direction that will *oppose* the change of current flow. When, for example, a coil is suddenly connected to a battery, the induced voltage acts to oppose the battery voltage. When the coil is disconnected from the battery, the induced voltage will aid the battery voltage.

Both of these effects, which are illustrated in Figure 6.16, are momentary, they last only for the brief period in which current flow through the coil is changing. If the current through the coil is an *alternating* current (which is continually changing) an *induced a.c. voltage* will always be present. This induced voltage is also alternating, and it acts to oppose the flow of current, making the coil behave as if it possessed greater resistance to a.c. than for d.c.

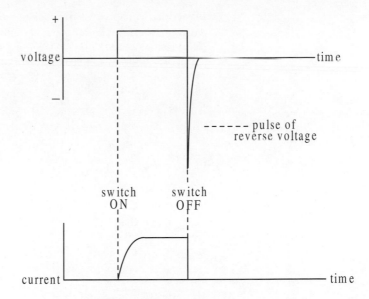

Figure 6.16   The effect of inductance when current is switched ON and OFF

Because it is the coil itself that induces the opposing voltage, this opposing voltage is called an e.m.f. of *self induction* or *back e.m.f.* The size of the induced voltage depends on the rate at which the flow of current changes, and on the shape and size of the coil. These *geometrical factors* of shape and size, are measured by the quantity called *self-inductance*, which is measured in units of henries (H). A coil has an inductance of 1 henry when a rate of change of current of one ampere per second causes an induced voltage of 1 volt.

In practical work, the smaller units of milli-Henries (mH) and micro-Henries (μH) are more useful. A few centimetres of straight wire will have an inductance of less than 1 μH. A coil used to tune a medium-wave radio will only have an inductance of some 650 μH. An iron cored *choke* which is intended to act as a high impedance to audio signals might typically have an inductance of 50 mH. Only a very large coil wound on a massive laminated iron core would have an inductance which needed to be measured in units of henries.

Any coil or inductor will provide a reactance to signal currents. This reactance, measured in ohms and abbreviated as X_L, is defined in a similar way to the reactance of a capacitor (see Chapter 5) as the value of:

$$X_L = \frac{\text{signal voltage across the coil}}{\text{signal current through the coil}}$$

Figure 6.17 shows how reactance depends on (a) inductance and (b) frequency. The reactance value of a coil is proportional to both inductance and frequency. A coil can therefore have almost zero resistance to d.c., but a large reactance to a.c. As a formula, the reactance of an inductor is as

$$X_L = 2\pi fL$$

shown, left. The value of reactance is in ohms if the inductance (L) is in units of Henries (H) and the frequency (f) in Hertz.

We use inductors in electronics to provide low resistance to d.c. and high reactance to a.c., almost the reverse of the capacitor action. Note however, that a capacitor will block d.c. completely, but no inductor can block a.c. completely.

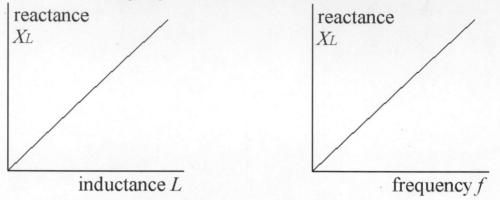

Figure 6.17   How inductive reactance varies with (a) value of inductance and (b) signal frequency

Question 6.6

What is the reactance of a 0.5 H coil at 100 Hz?

Transformer action

The changing flux around a coil that carries a changing current will also induce a voltage in another coil. This is the principle of the *transformer*, a device which is used extensively in electricity and electronics. A simple transformer, pictured in Figure 6.18(b), consists of two coils that are wound on a common magnetic core. When the current through one coil (the primary winding) changes, a voltage is induced in the other winding (the secondary winding).

Remember: a single inductor has self-inductance. When inductors are wound on the same core, or close together, each winding still has self-inductance, and there will also be mutual inductance between windings. The *mutual inductance* that exists between the windings of a transformer means that a changing current passed through the secondary winding will cause a voltage to be induced in the primary winding. The construction of a transformer will be more easily understood if you look at the cross-section of a strip of transformer core and its windings illustrated in Figure 6.18(c). The primary and secondary windings of a transformer are arranged to be as close to each other as possible.

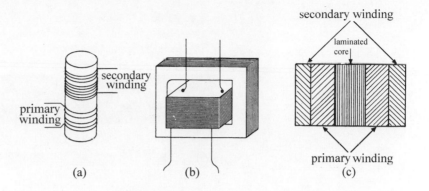

Figure 6.18 (a) Transformer principle; (b) a typical small transformer, (c) cross-section of windings on a transformer core

Activity 6.6

Connect a 100 mH choke in series with a resistor of about 5 K across the output of an audio signal generator. Apply a constant level of signal at frequencies ranging from 100 Hz to 10 kHz and measure the signal voltage developed across the choke, using an a.c. voltmeter. Plot the output voltages against frequency. Explain why this graph has a similar shape to those shown in Figure 6.17.

Actions

One of the principal actions of a transformer is to pass signals from one circuit to another without any connection other than the magnetic flux existing between the two circuits. The transformer therefore passes a.c. signals but blocks d.c. A transformer can also achieve a *step-up* or *step-down* of alternating voltage depending on the ratio of the number of turns of wire in each of its two windings. For a given a.c. voltage across the primary winding (V_p), the voltage across the secondary winding will be

$$V_s = V_p \frac{N_s}{N_p}$$

where N_s is the number of turns in the secondary winding and N_p is the number of turns in the primary winding of the transformer. If the transformer is perfect, meaning that there is no power loss between primary and secondary, the current ratio I_p/I_s will be equal to N_s/N_p, the turns ratio.

The construction of a transformer core depends on the frequency range of the signals which the transformer is designed to handle. For the lowest audio frequencies, a large core of soft magnetic material in the form of thin laminations is enough At higher frequencies, ferrite dust cores must be used to reduce as far as possible the magnetic losses (eddy-current losses) which arise in metal cores when high frequency signals pass through them. At the highest frequencies for which coils are used, no metal core material is acceptable. Coils have either to be wound on plastic formers, or they are made self-supporting. For UHF (and higher) frequencies, short parallel metal strips carry out transformer action.

Question 6.7

A step-down transformer uses a 4:1 step-down ratio. If the primary voltage is 240, what output can be expected at the secondary?

The autotransformer

The autotransformer uses a single winding to act as both primary and secondary of the device. As will be seen from the circuit symbol in Figure 6.19, it amounts to a tapped inductor. The autotransformer behaves exactly as does the transformer just described, except that it provides no d.c. isolation between its primary and secondary windings.

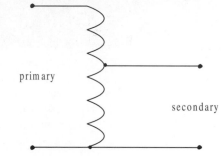

Figure 6.19   The autotransformer symbol

The ratios of voltage, current and resistance (impedance) in the autotransformer are identical to those in the ordinary transformer, and the ratio of turns is worked out in exactly the same way. Many autotransformers are used for stepping down voltage, so that the primary number of turns is the total number of turns on the core, and the secondary number is the small amount taken from one end of the winding to the tap point.

A commonly used type of autotransformer, called a Variac™, has a sliding tapping point which can be continuously varied to provide an output (secondary) range of zero to some 130 per cent of the nominal primary voltage. It is useful in circuits in which the voltage needs to be set precisely to a known value.

Activity 6.7

Connect the output of an audio signal generator set to a frequency of 1 kHz to the primary winding of a small audio transformer having a primary/secondary turns ratio of 5:1. Terminate the secondary winding with a resistance load of 3 to 5 ohms. Using a multirange meter, measure both the primary and secondary voltages. Break the primary and secondary circuits to allow the addition of the meter in series so that the currents can be measured. Compare the resulting voltage and current ratios.

Magnetic hysteresis

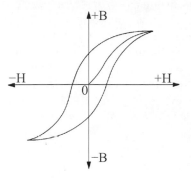

When a material is magnetized and demagnetized, its behaviour is unexpected. Suppose we wind a coil of wire around a magnetic material, and change the current through the coil. Starting from zero, for each reading of current we measure the magnetic strength of the core material, and we plot values for currents rising and falling in one direction and also rising and falling in the reverse direction. The shape of this graph is a type of curve called a *hysteresis loop*, illustrated, left.

There are several important points that we can learn from this graph:

After one cycle, there is a value of magnetic strength even when no current flows. This is the permanent (or remanent) magnetism for the material. This will be large for a hard magnetic material, small for a magnetically soft material.

The graph follows a different path for reducing current compared to the path taken when current is increasing.

The magnetic strength in either direction reaches a maximum, and even passing large additional current will not alter this value.

You have to pass a current permanently to keep the material demagnetized.

In practice, a magnetic material can be completely demagnetized if a.c. is passed through the coil, and the material is slowly taken away from the coil. This makes the material cycle through the hysteresis loop thousands of times, with the peak of the loop getting smaller on each cycle until the magnetic strength is almost zero.

Magnetic tape

Magnetic tape uses the principle of hard magnetic material. Early tape recorders in 1898 used steel tape, but it was not until 1941 that the BASF company hit on the idea of coating a thin plastic tape with magnetic powder, using iron oxide. The recording head is made from a soft magnetic material with low hysteresis so that its magnetic flux will follow the fluctuations of a signal current through the winding. The same head is also used for replaying tape, using the varying flux from the moving tape to induce currents in the head.

Because iron oxide is a hard magnetic material, it can be magnetized by a strong magnetic field close to the tape, and this magnetism will be retained when the tape is removed from the field. When the magnetized tape is drawn past a coil, variations of magnetism in the tape will induce varying voltages in the coil, so providing the replay action. See Chapter 16 for a more complete description of tape recorder principles.

Practical inductors

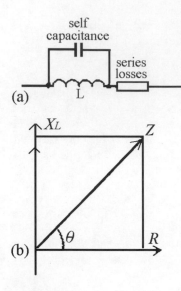

(a)

(b)

C-I-V-I-L

C – I before **V**

V before **I** in **L**

Because it is made from a coil of wire, any practical inductor will possess some inherent resistance. Though small by design, this will dissipate energy in the form of heat. In addition, the very small volts drop across each adjacent turn creates a parallel self capacitance. This in turn creates a self resonant frequency for the device which leads to further losses. Fortunately, the self capacitance is so small that it can be ignored in most circuit design cases. The equivalent circuit of a practical inductor is shown in the drawing (a), left.

The figure (b) shows the phasor/vector diagram relationship between resistance (R), inductive reactance (X_L) and total impedance (Z). This indicates the phase angle θ. The phase angle found in this way is the phase between the applied ac voltage (V) and the resulting current (I). If the device had been perfect, the phase angle θ between V and I would be 90° with V leading I. However in practice, the inherent resistive component reduces this angle. The resulting total impedance (Z) of the device, including the resistance R is given by $Z = \sqrt{X_L^2 + R^2}$ ohms

The phase angle is related to the Quality (Q) factor for the inductor, where $Q = X_L/R$. Hence the lower the value of R then the higher the Q factor. Q is also the ratio between the energy stored in the inductor by the magnetic field to that dissipated by the resistor per cycle of a.c. current.

The word **CIVIL** is often used as an aid to memory in regard to phase shifts. Read it as **C**apacitive circuit, **I** leads **V** but in an inductive circuit **V** leads **I**.

Answers to questions

6.1   The moisture will permit leakage current which might be greater than the signal current through the capacitor.

6.2   Many transmitters use high voltages and electrolytics cannot reliably be used.

6.3   470 microseconds.

6.4   338 ohms.

6.5   The magnetism of a solenoid can be switched on or off; a permanent magnet does not need power to maintain its magnetism.

6.6   314 ohms.

6.7   60 V.

7

A.c. mains supply and practical electronics methods

Mains supplies

The mains supply to a house is a.c. at around 240 V, 50 Hz. This has been transformed down from the very high voltages used for distribution, but factories normally use a higher-voltage supply of around 415 V a.c.

This arises because electricity is distributed in three phases using a set of four cables. The fourth cable is earthed at the supply transformer station. This arrangement provides for 415 V three-phase power for industry, along with the 240 V supply between each phase and earth. The house supply will be taken in this way, and the number of houses using each phase will be balanced so that approximately the same current has to be supplied by each phase wire.

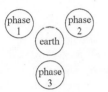

The wiring in a house or factory uses mains cable. For a house this consists of stranded copper made into three cores, each insulated and with an outer insulated PVC cover for protection. House wiring uses a ring construction, so that each wiring socket is connected in a ring, taking current from all of the cables in the circuit. This scheme has been used for some fifty years, and it makes fusing and cabling simpler, as well as requiring less cable. Wiring in a factory is likely to use either three or four conductors in a cable, and it may require cables that are insulated with fireproof minerals and with a metal outer cover.

The conventional house sockets use live, neutral and earth connectors. The live connection provides the full 240 V a.c., and the neutral connection is used to complete the circuit. Any equipment that is plugged in will therefore be connected between the live and the neutral. The earth connection is taken to buried metalwork in the house, and is used to ensure that equipment with a metal casing can be used safely, because the casing will be connected through the three-pin plug, to earth. Wiring regulations specify that *protective multiple earthing* (PME) will be used, so that all metal pipes in a house must be connected using large capacity cable to the same earth.

Mains plugs

The greatest hazard in most electronic servicing is the mains supply which in the UK and in most of Europe is at 240 V a.c. Three connections are made at the usual UK domestic supply socket, labelled live (L), neutral (N) and earth (E) respectively. The live contact at 240 V can pass current to either of the other two. The neutral connection provides the normal return path for current with the earth connection used as an emergency path for returning current in the event of a fault. Earth leakage circuit breakers (ELCBs) work by detecting any small current through the earth line and using this to operate a relay that will open the live connection, cutting off

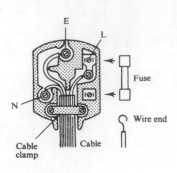

the supply. Connection to domestic supplies is made by a three-pin plug, shown opened in the illustration. This plug is designed so that shutters within the socket are raised only when the plug is correctly inserted.

Table 7.1   Colour codes for flexible wiring

	Live	Neutral	Earth
Modern	Brown	Blue	Green/Yellow
Old colours	Red	Black	Green

Table 7.1 lists the colour coding of the wire connections to the plug. Though it is many years since the coding colours were changed, older equipment can still be found bearing the older colours. Note that these are the colour codes for flexible wiring, and internal house wiring still uses the older Red, Black, Green scheme.

Particular care should be taken in working on cables which have been colour coded to foreign standards, though by the mid 1990s all European equipment should be using the same coding. The older colours are still used in internal domestic wiring to sockets. A plug must be wired so that:

- The cable is firmly held and clamped without damaging the insulation or the conductors.

- All connections are tight with no loose strands of wire. The ends of the cable can be coated with solder to prevent loose strands from separating, but the soldered end should not be used for clamping because the wire is more brittle and will loosen off after some time.

The wires should be cut so that the live lead will break and pull free before the earth lead if there is excessive force on the cable. The design of some plugs can make this very difficult, and you should select plugs that permit the use of a short live lead.

A fuse of the correct rating must be used. The standard fuse ratings for domestic equipment are 3 A, colour coded red, and 13 A, colour coded brown. Most domestic electronic equipment can use the 3 A rated fuses, and the few items that require a larger fuse should preferably be used with a (non-standard) 7 A fuse rather than the 13 A type, because the appliance cables for such equipment are seldom rated for 13 A.

Table 7.2   Fuse ratings

Value	Colour	Applications
3 A	Red	All domestic electronic equipment; test gear
13 A	Brown	Heaters and kettles

The cable should be clamped where it enters the equipment, or alternatively a plug and socket of the standard type can be used so that the mains lead consists of a domestic plug at one end and a *Eurosocket* (never a plug) at the other. The Eurosocket is also referred to as an *IEC connector*.

Note: Non-domestic equipment often makes use of standard domestic plugs, but where higher power electronic equipment is in use the plugs and socket will generally be of types designed for higher voltages and current, often for 3 phase 440 V a.c. In some countries, flat two-pin plugs and sockets are in use, with no earth provision except for cookers and washing machines, though by the year 2000 uniform standards should have prevailed in Europe, certainly for new buildings.

The circuit should also include:

- A fuse whose rating matches the consumption of the equipment. This fuse may have blowing characteristics that differ from those of the fuse in the plug. It may, for example, be a fast-blowing type which will blow when submitted to a brief overload, or it may be of the slow-blow type that will withstand a mild overload for a period of several minutes.

- A double-pole switch that breaks both live and neutral lines. The earth line must never be broken by a switch.

- A mains warning light or indicator which is connected between the live and neutral lines.

All these items should be checked as part of any servicing operation, on a routine basis. As far as possible all testing should be done on equipment that is disconnected and switched off. The absence of a pilot light or the fact that a switch is in the OFF position should never be relied on. Mains powered equipment in particular should be completely isolated by unplugging from the mains. If the equipment is, like most non-domestic equipment, permanently wired then the fuses in the supply line must be removed before the covers are taken from the equipment. Many pieces of industrial electronics equipment have safety switches built into the covers so that the mains supply is switched off at more than one point when the covers are removed.

The UK has used the three-pin plug with rectangular pins for some time, but in the lifetime of most of us these may be replaced by the pattern used on the Continent. The Continental pattern is that most plugs are two-pin, with three used only for a few items such as electric fires and cookers. There is no fuse in the plug, but an earth-leakage circuit breaker (*ELCB*) is incorporated into the socket to cut off current if any leakage to earth is detected – similar contact breakers are available in the UK for use with power tools out of doors.

Special regulations apply to electrical tools and other equipment that can be described as *double-insulated*. No earth connection is required for such equipment because the metal parts that are connected to the supply (such as the motor of a power tool) are insulated from the casing, even if the casing is metal. Equipment with plastic casings are normally of the double-insulated type of construction.

The current rating of domestic plugs in the UK is a maximum of 13 A, but for electronic equipment, fusing at 3 A is more common. Despite

fuse
wire

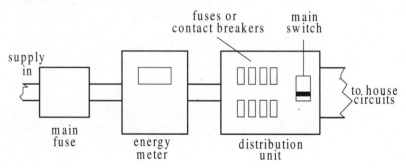

metal
caps

attempts to standardize only the 13 A and 3 A fuses, you can buy intermediate values such as 5 A and 7 A.

Fuses are pieces of thin wire in an insulating container, and the principle is that when excessive current flows the fuse will melt, and so cut off the current. A fuse will not melt immediately, and at the rated current it may take several minutes to blow. A fuse provides protection from a disastrous short circuit that causes a large current to flow, but it provides no protection to humans or to some types of electronic equipment. Contact breakers, which can break the circuit on a precise and very small value of excess current, are to an increasing extent replacing fuses as a method of protection.

The wiring of a house, Figure 7.1, is centred at the *distribution unit*. The cable from the local transformer, which may be overhead or underground, is connected to a *main fuse* which is sealed and can legally be opened only by an employee of the electricity supply company. From the main fuse, cables connect to the *energy meter*. This is what we usually call the electricity meter, and it measures both current and time, so that the reading is in joules (in fact, kilojoules, kJ) so that electricity use can be costed. This is also a sealed unit, and modern types often incorporate a switching circuit that allows the cost per unit to change at preset times. This allows for the use of a cheaper rate at night, typically between midnight and 0700 in the morning.

From the energy meter, cables connect to the distribution unit. This contains a main switch that will isolate all the circuits in the house, and a set of fuses or contact breakers that are used to supply the individual ring circuits.

Figure 7.1   The main fuse, energy meter and distribution unit wiring

All the lighting and power circuits of the house are joined in to this distribution unit, which is provided with fuses or, more commonly now, contact breakers. Contact breakers are more sensitive than fuses, so that it is quite common to have a contact breaker break when, for example, an electric bulb fails. The contact breaker can be reset after the bulb has been replaced. A disadvantage of this sensitivity is that some equipment, notably computers, can be badly affected by a sudden power failure that might be caused by a minor fault. A modern distribution unit will use one residual current device (RCD) which will break the supply if there is any current

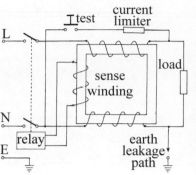

leaking to earth. In addition, it will use miniature contact breakers (MCB) in place of fuses.

RCD units can be bought separately, particularly for electrical equipment used outdoors. They should be used to protect test equipment, but *not* in a cable that is connected to or from an isolating transformer.

As implied above, MCB/RCD devices are intended to prevent electrical shock which may originate from the live metalwork of a system under fault conditions. They are designed around a special transformer core that carries three windings, see diagram, left. Two are connected in series with the Live (L) and Neutral (N) terminals of the supply. These are wound and connected in such a manner as to create equal but opposing magnetic fluxes in the core when the system is working correctly.

Under this condition, the third sense winding does not generate an output voltage. If a fault develops such that some of the load current flows to earth, the current in the N line will fall whilst that in the L line rises. As a result of this unbalance, a voltage will be induced into the sense winding which will operate the relay and break both of the L and N connections to the circuit. A circuit test button is usually fitted to the RCD devices which when pressed causes a limited current to flow and set up the necessary unbalance to trip the relay. Once the cause of the extra earth leakage has been cleared, the device can be reset.

Similar devices are available for use on 3-phase industrial power supplies, but these are usually arranged so that an earth fault on one phase will shut down all three.

MCB/RCD devices are rated by the continuous current that can be supported by the protection contacts, ranging from values of 5 A to 125 A or even higher for special applications. The tripping current is also specified and this is typically within the range of 5 mA to 2 A, with 5, 10 and 30 mA being commonly used on domestic power supplies. The operating frequency is also specified for application on either 50 Hz or 60 Hz mains supplies.

To replace a fuse, the main switch must be turned off, and the faulty fuse removed. At one time, a fuse consisted of fuse-wire connected into a ceramic holder, but it is more usual now to have tubular fuses whose action is more predictable. The older type is dealt with by unscrewing the severed wire and replacing with fuse wire of the correct rating. The more modern type uses a replacement tubular fuse, and spares of each required rating should be kept at or near the distribution box. Contact breakers (MCBs) do not need to be removed, nor is it necessary to turn off the main switch. To reset a contact breaker you simply press the button or flick the switch of the unit that has broken the supply.

The cables that are used to connect sockets to the distribution unit are normally buried in the walls of a house, though eventually builders may start to take them inside hollow skirtings to make it much easier to add sockets. The connection between plug and an electrical appliance is made by way of flex, which is a form of cable that uses fine strands of copper to make it bend easily.

Flex is easily damaged, and you should avoid stepping on it, working with sharp instruments near a flex, bending or coiling the flex too tightly, and running flex over hot surfaces. All flexes, particularly those that are used for high-current supplies, should be inspected at intervals, and their connections checked. This applies particularly to power tools.

Signal cables are also very vulnerable. These are often of the screened type, with a copper mesh surrounding the insulated wires of the cable, and this mesh needs to be earthed, usually by way of the connector. Signals cables should be treated with the same respect as flex.

Question 7.1

Your electricity supply has failed and the fuses in the consumer unit are all intact. What should you do?

Constructing electronic circuits

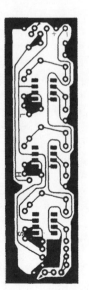

The methods that are used industrially to construct circuit boards in large numbers are quite different from the methods that have to be used for experimental or one-off circuits. One factor that is common to all constructional methods, however, is that the circuit diagram is a way of showing connections which does not give any indication of how the components can be physically arranged on a board.

The main difference between circuit diagrams and layout diagrams is that on a layout diagram any crossing of connecting leads has to be avoided. A component such as a resistor or capacitor can cross a circuit connecting track because on a conventional printed circuit board (PCB) the component will be on the opposite side of the board from the track (but see surface mounting, later in this chapter). The simplest circuits to lay out are discrete transistor amplifiers; the most difficult are digital circuits in which each chip contains a large number of separate devices.

Small-scale circuits can be laid out manually, using cardboard cut-outs of the components, with connections and internal circuits marked, on a large sheet of tracing paper or on transparent plastic which has been marked with a pattern of dots at 0.1 inch (2.5 mm) centres. The designing is done looking at the component side of the board, and will start by roughing out a practicable layout which does not require any track crossovers. At this stage, it is important to show any interconnecting points, using edge connections or fixed sockets, because it must be possible to take leads to these connectors without crossovers. This can be quite difficult when the connection pattern is fixed in advance, as for example when a standard form of connection like a stereo DIN socket or a Centronics printer socket is to be used.

This layout can then be improved, with particular attention paid to durability and servicing.

- The positions of presets and other adjustments will have to be arranged, along with points where test voltages can be measured, so that servicing and adjustment will be comparatively easy.

- Signal lines may have to be re-routed to avoid having lines running parallel for more than a few millimetres (because of stray capacitances), or to keep high impedance connections away from power supply leads.

- Some tracks that carry RF may need to be screened, so that earthed tracks must be provided to which metal screens can be soldered.

- Components which will run hot, such as high wattage resistors, will have to be mounted clear of other components so that they do not cause breakdown because of overheating in semiconductors or capacitors. One way of achieving this is to use long leads for these components so that they stand well clear of the board, but this is not always a feasible solution if several boards have to be mounted near to each other.

More extensive circuits can be planned by computer, using software such as the EASY-PC program, so that the PCB pattern is printed out after details of each component and each join in the circuit have been typed into the computer. Even computer-produced layouts, however, may have to be manually adjusted to avoid having unwanted stray capacitances appearing between components. Either method will have to take account of the physical differences that can exist between similar components, such as length of tubular capacitors and the difference between axial and radial lead positions.

One-off and experimental circuits can be constructed on matrix boards (also known as stripboards, see later), and the layout that is used on these boards can be used also as a pattern for manufacturing PCBs for mass production. In either case, a board material must be used which will be strong and heat resistant, with good electrical insulation. The choice will usually be between plastic impregnated glass fibre board, or SRBP (synthetic resin bonded paper), though some special purpose circuits may have to be laid out on ceramic (like porcelain) or vitreous (like glass) materials in order to cope with high temperature use and flame-proofing requirements.

Whatever type of board is used will be covered with copper to act as the connections between components. On a matrix board, the copper will be laid out in strips, usually 0.1" apart and drilled with holes for leadout wires also at 0.1". Boards for larger scale production are un-drilled and completely covered with copper, which will then be etched away into the pattern of connections that is needed. For mass production, the processes are completely automated, and the assembly of the components on to the board and subsequent soldering will also be totally automatic. The components are then inserted into holes drilled in the mounting pads to which the leads are soldered, usually in a solder bath so that all connections are soldered in one action.

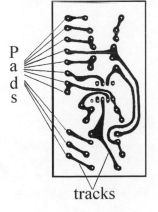

P a d s

tracks

Many commercial PCBs, particularly for computer or other digital applications, are double-sided, with tracks on the component side as well as on the conventional track side. Where connections are needed between sides, plated-through holes are used. These are holes which have copper on each side and which have been electroplated with copper so that the holes have become partly filled, making a copper contact between the sides. These connections are strengthened when the board is soldered.

The use of double-sided board is particularly important for digital circuits where a single-sided board presents difficulties because of the need to cross leads. The use of a well designed double-sided board can solve these problems, but care needs to be taken over capacitances between tracks that are on opposite sides of the board.

The main board (or *motherboard*) of a computer normally uses more than two layers of conductors, and such boards must *never* be drilled or cut. Considerable care is also needed in soldering to or de-soldering components from such a board.

Surface mounting

Surface mounting technology (SMT) has been in use for some time, certainly since 1977, but only recently has its use become widespread. Components for surface mounting use flat tabs in place of wire leads, and because these tabs can be short the inductance of the leads is greatly reduced. The tabs are soldered directly to pads formed on to the board, so that there are always tracks on the component side of the board. Most SMT boards are two-sided, so that tracks also exist on the other side of the board.

The use of SMT results in manufacturers being able to offer components that are physically smaller, but with connections that dissipate heat more readily, are mechanically stronger and have lower electrical resistance and lower self-inductance. Some components can be made so small that it is impossible to mark a value or a code number on to them. This presents no problems for automated assembly, since the packet need only be inserted into the correct hopper in the assembly machine, but considerable care needs to be taken when replacing such components, which should be kept in their packing until they are soldered into place. Machine assembly of SMT components is followed by automatic soldering processes, which nowadays usually involve the use of solder paint (which also retains components in place until they are soldered) and heating by blowing hot nitrogen gas over the board. Solder baths are still used, but the hot gas method causes less mechanical disturbance and can also allow heat sensitive components to be shielded.

Considerable care is needed for hand soldering and unsoldering SMT components. A pair of tweezers can be used to grip the component, but it is better to use a holding arm with a miniature clamp, so that both hands can be free. The problem is that the soldering pads and the component itself can be so small that it is difficult to ensure that a component is in the correct place – it can even be difficult to distinguish an SMT component from a solder pad. Desoldering presents equal difficulties it is difficult to ensure that the correct component is being desoldered, and almost impossible to identify the component after removal. A defective SMT component should be put into a 'rejects' bin immediately after removal.

Small scale circuit production

The simplest form of construction for a one-off circuit is the matrix stripboard. This can be obtained in a wide range of sizes, up to 119 x 455 mm (4.7" x 18" approx.). Tracks can be cut by a 'spot cutter' tool which can be used in a hand or electric drill, this allowing components like DIL ICs to be mounted without shorting the pin connections, see Figure 7.2. Once these cuts have been made, the components can be soldered on to the board and the circuit tested. Arrangement of components can be tested in advance by using a solderless breadboard, which allows components to be inserted and held by spring clips, using a layout which is essentially the same as for matrix stripboards.

soldered points

tracks cut

Figure 7.2   A matrix stripboard circuit

An alternative is the one-off PCB. The pattern of the circuit tracks has to be drawn on a piece of copper laminated board, using a felt tipped pen which contains etch-resistant ink. An alternative is to work from a transparency of the pattern, using light sensitive etch resistive material which is then 'developed' in a sodium hydroxide solution. When the ink or other etch resist is dry, the board is etched in a ferric chloride bath (acid hazard – wear goggles, gloves and an apron) until all the unwanted copper has been removed.

In a mass production process, the pattern of etch resist is placed on to the copper by a silk-screen printing process. The copper is then etched in baths which are maintained at a constant high temperature, and the boards are washed thoroughly in water, followed by demineralized (soft) water, and then finally in alcohol so as to make drying more rapid. For the handmade etched board, all traces of the resist material and the etching solution have to be removed by washing and scrubbing with wire wool.

The board can then be drilled, using a 1 mm drill, and the components inserted. The final action is soldering, using an iron with a small tip. The boards are heat resistant, not heat-proof, so that soldering should be done fairly quickly, never keeping the iron in contact with the copper for too long. Excessive heating will loosen the copper from the plastic board, or burn the board, and if the copper has been cleaned correctly and all components leads are equally clean, soldering should be very rapid. Chemical tinning solutions can be used to treat the copper of the board so that soldering can be even more rapid. Remember that excessive heat will not only damage the board but also the more susceptible components like semiconductors and capacitors. The time needed to obtain a good soldered joint should not exceed a few seconds.

Laying out a printed circuit

A straightforward circuit can be laid out on matrix board by following a simple set of rules. First, the circuit junctions must be found. A circuit junction is a place where components are connected either to one another or to wires leading from or to the PCB. On a matrix or PCB every circuit junction or *node* is represented by a separate strip or area of copper. Figure 7.3 shows some examples of circuit junctions (ringed), such as would be found in simple circuits.

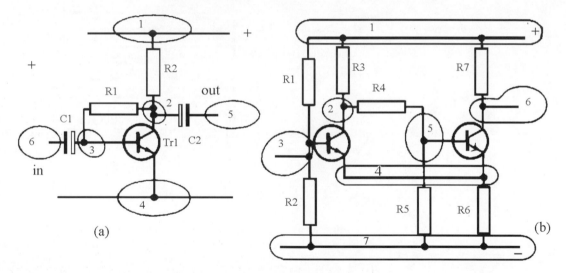

Figure 7.3   Marking and numbering circuit junctions on a diagram

Once they have been identified and ringed with a pencil on the circuit diagram, the circuit junctions are numbered, starting with 1 for the junction which takes in the (+) supply line. When the circuit contains a transistor, it is an advantage to have the circuit junctions for the three transistor connections numbered consecutively i.e. 3, 4, 5 or 8, 9, 10 so that transistors with short leadout wires can be used.

When the junctions have been numbered, the circuit can be built on to matrix boards with numbered strips. Some types of matrix board are numbered during manufacture. Others can be numbered by gumming paper strips on to each side at the ends of the tracks and numbering the tracks on the paper.

- An alternative is to coat each side of the board at the ends with one of the white fluid materials which are used for correcting typing errors (sold under such names as Snopaque and TippEx). These fluids dry very rapidly, leaving a matte white surface on which it is easy to write.

Once the numbering is complete, the components are soldered in place to the tracks whose numbers are shown in the circuit diagram, prepared as already described. Taking the example of Figure 7.3(a), R_1 is soldered between Tracks 2 and 3, R_2 between Tracks 1 and 2, C_1 between Tracks 6 and 3 with the positive end on Track 3, and C_2 between Tracks 2 and 5 with the positive end on Track 2. The transistor is connected with its emitter to Track 4, its base to Track 3, and its collector to Track 2. The supply is taken to Tracks 1 (+) and 4 (−), the input signal to Track 6 and the output signal from Track 5.

There is a variation of this method which is particularly suitable for symmetrical circuits such as the multivibrator circuit illustrated in Figure 7.3(b). A central track is used for the negative supply line, and two separate tracks linked by a wire soldered to each are used for the positive line.

The layout of a circuit on to copper laminate board should start by making a drawing on tracing paper or transparent film. Components, or cardboard cut-outs can then be placed on the drawing to show sizes, and to mark in the mounting pads to which the leads will be soldered. This has been done with the drawing representing the component side of the board, but the tracks can now be drawn in as they will exist on the copper side this means that the actual appearance of the component side will be the mirror image of your layout.

The drawing will then have to be transferred to the copper. This can be done manually, using an etch-resistant ink as described earlier, or by photographic methods. The board can then be etched, thoroughly cleaned, drilled and then the components mounted and soldered into place.

Question 7.2

What particular section of the HSE regulations apply to the preparation of printed circuit boards?

Soldering and desoldering

Soldering is a way of joining metals by melting a metal alloy of lead and tin on to the other metals. Metals that are to be joined in this way must be compatible (aluminium needs special treatment), thoroughly clean, and should be mechanically joined so that the soldered joint is not expected to provide support as well as electrical connection unless the amount of support is minor. The main obstacle to good soldering is dirt, and thorough

cleaning is the most important part of good soldering, whether by hand or by automatic solder bath methods.

Soldering irons (despite the name, the business end is made from copper) are available in various sizes, and for electronics use, a 15 W type along with a 25 W type will cover most of the circuit requirements, though a larger iron is useful for some specialized tasks. For some heavier work, a fast-heating low-voltage iron (a solder-gun) can be useful, and for some types of servicing a battery operated 6 W iron may be needed. Some irons can have de-soldering attachments clipped on, and a set of these will be needed unless other de-soldering methods are used. The de-soldering provision should be a solder pump, described later in this chapter.

The tips of bits should be kept clean and in their correct shapes. The copper bits can dissolve quite rapidly in some types of solder, though many solder formulations contain some copper to inhibit this action. A supply of spare bits, and a spare element for each type of iron, should be stocked. Scale should be removed from the main body of the iron at intervals, and the earth resistance checked, because an earthed iron is safer to use, particularly for modern types of ICs.

Solder should be of the flux cored variety, but a tin of flux should be kept handy in case of large soldering jobs. Having a separate tin of flux also makes it much easier to 'tin' (coat with solder) large surfaces of metal using non-cored solder and either a large iron or a miniature blowlamp.

To solder a joint effectively, the metals should first be thoroughly cleaned, and in some cases it is preferable to tin both. They should then be mechanically connected by twisting wire, bending tags etc., so that the joined metals will not depend on the solder for mechanical strength. The iron is allowed to heat thoroughly until it will instantly melt a piece of flux-cored solder held to the tip. The tip should be kept clean just before use by wiping it with a slightly damp cloth. Never flick excess solder from the tip of an iron.

The iron is held against the metals that are to be joined for a few seconds so as to heat them, and then the solder is applied to where the tip of the iron contacts the joint. The solder will melt and spread over the joint, and as soon as the solder makes a smooth joint, the iron can be removed. Never hold the iron in place longer than is needed to allow the solder to become smooth like this, as excessive heat will burn off the flux, allow the solder to oxidize in the air, and ruin the electrical and mechanical soundness of the joint. Figure 7.4, overleaf, shows the difference between correct and incorrect solder flow.

A few types of metals are notoriously difficult to solder, including spring steel and some of the nickel alloys that are used for heating elements. These need scrupulous cleaning, often with acid washes, and should be tinned beforehand.

Large pieces of metal should be tinned using a small blowlamp. Heat sensitive components require rather different treatment. The leads of transistors should be gripped by pliers so as to act as a heat shunt, preventing the heat of the iron from reaching the casing of the transistor.

ICs cannot be gripped in this way, but it is usually possible to clamp a 'bulldog' type of clip over the roots of the pins. Particular care should be taken to avoid excessive heating on any electronic components, but semiconductors and some capacitors are by far the most susceptible to damage.

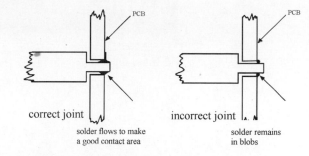

correct joint

solder flows to make
a good contact area

incorrect joint

solder remains
in blobs

Figure 7.4   Correct and incorrect soldering

Desoldering should preferably be done with a 'solder pump' or sucker. This is a spring loaded suction pump with a PTFE nozzle. The plunger is depressed, expelling the air and the nozzle held to the joint which is to be desoldered. A hot soldering iron is applied and when the solder is fluid, the plunger is released by pressing the trigger button. This action sucks the molten solder into the pump body, from where it can be removed at intervals.

It cannot be emphasised too strongly that the skills of soldering must be thoroughly learned at an early stage. An electronics technician without this ability will be an encumbrance to any servicing workshop. Poor soldering technique will be noticeable in any practical test, and it is easier to learn the correct method from the start than to try to forget bad habits.

The traditional solder alloy is 60 per cent tin and 40 per cent lead. Because lead is considered to be an environmental hazard (despite the millions of healthy people who grew up in houses equipped with lead water pipes), alloys that use no lead content are now considered desirable. Typically, such lead-free alloys use silver, bismuth, or both. Replacement of lead solder has to be done carefully, because the replacement alloy must match the traditional type closely, particularly in respects such as melting temperature, ability to wet and coat copper, good coverage of copper, lifetime, matched expansion, etc.

The types of flux, traditionally rosin or a borax solution or paste, that are used are also now subject to change because of the choking fumes that traditional flux materials can cause in a confined area. Modern flux materials emit much less smoke during soldering and are easily removed from the work after soldering is complete.

For intensive soldering work or for soldering that involves surface-mounted components (SMCs), a *solder station* is often a preferred tool. A solder station will consist of a bench fitted with air-extraction, and with a range of soldering and desoldering tools with a selection of bit sizes, jigs and holders for circuit boards, as well as hot-gas tools for use with SMCs.

Activity 7.1

Identify the circuit junctions in the circuits shown in Figure 7.4(a) and (b). Draw them in, and check them carefully. Build the circuits either on matrix board or by plugging the component wires into 'solderless breadboards', and apply the voltages shown. (Tr_1, and Tr_2 are general purpose NPN transistors.)

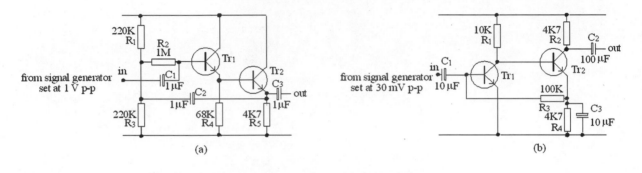

(a)   (b)

Figure 7.4   Circuits for Activity 7.1

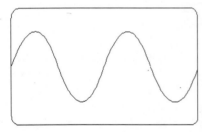

The circuits, when operating correctly with the specified input, will each give an output waveform which will appear on the screen of an oscilloscope as shown, left. If the waveform is distorted check first that the input amplitude is correct, before looking for incorrect voltage levels in the circuit or faulty components.

Answers to questions

7.1   Telephone the electricity supplier, because you cannot replace the main fuse, and the problem may be a power cut.

7.2   Control of Harmful Substances (COSH).

Unit 3

1. Demonstrate an understanding of semiconductor diodes and d.c. power supplies and apply this knowledge safely in a practical situation.

2. Demonstrate an understanding of semiconductor active devices and apply this knowledge safely in a practical situation.

Note: This unit calls for practical work in assembling and testing transistor and IC amplifier circuits. This work is dealt with in Chapters 7 and 12 because it calls for knowledge of waveforms that has not yet been established.

8

Semiconductor diodes

Semiconductors

Semiconductors are not simply materials whose resistivity is somewhere between that of a conductor and that of an insulator. Certainly a pure semiconductor will have a resistivity value that is not so high as that of an insulator, but it certainly does not approach the low value that we would expect of a conductor. The two items that make us class a material as a semiconductor are the effect of temperature and the effect of impurity, and both of these effects are closely related.

Suppose, for example, that we have a specimen of pure silicon. Its resistivity is very high so that we would normally think of this material as an insulator. When the pure silicon is heated, however, its resistivity drops enormously. Though the drop is not enough to place hot silicon among the ranks of good conductors, the contrast with any other insulators is quite astonishing.

The effect of impurities is even more amazing. Even tiny traces of some impurities, one part per hundred million or so, will drastically change the resistivity of the material. It's because of this remarkable effect of impurity that we took so long to discover semiconductors – it was only in this century that we discovered methods of purifying elements like silicon to the extent that we could measure the resistivity of the pure material.

The main difference between a semiconductor element and any other element is that a semiconductor can have almost any value of resistivity that you like to give it. It is, in other words, a material that can be engineered to have the electrical characteristics that you want of it. The manipulation is done by adding very small quantities of other elements.

The first semiconducting materials to be used were the metallic elements germanium and silicon, and silicon is still the most-used semiconductor material. The compound called *gallium arsenide* is also a semiconductor, however, and it can be used to form diodes and transistors that will operate at very high frequencies, well into the microwave region. Gallium arsenide transistors are used in the LNB (low noise block) circuit that is part of a satellite TV aerial; this amplifies the feeble microwave signals and converts them to lower frequencies. Gallium arsenide is also used to form LEDs (see later, this chapter).

Electrons and holes

Even in the 19th century there was a suspicion that electric current through crystals of materials was not entirely caused by the movement of electrons. We know now that there are two ways that electric current can be carried in crystals – note that this applies *only* to *crystals*. Crystals are never perfect, and when a crystal of an almost pure material has been deliberately made impure (or *doped*), the crystal contains atoms of a different type. If these atoms possess more or fewer electrons in their outer layer than the normal atoms of the crystal, then the electrical characteristics will also change.

One way of changing the characteristics is to release more electrons. The other way is to release more holes. A *hole* is a part of a crystal that lacks an electron. Because of the structure of the crystal, a hole will move from atom to atom and when it does, it behaves just as if it were a particle with a positive charge. Within the crystal, the hole has a real existence, we can measure its charge and even a figure for its mass. The important difference is that the hole is a discontinuity in a crystal, it has no existence outside the crystal. The electron, by contrast, can be separated from the crystal, and can even move in a vacuum, as in a cathode-ray tube.

By adding impurities to pure semiconductor materials, then, we can create the value of resistivity we want, and we can also arrange it so that most of the current is carried either by electrons or by holes. If most of the current is carried by electrons, we call the material N-type, and if most of the current is carried by holes we call the material P-type. Note that it is only the current carriers inside the material that have positive or negative charges – the numbers of these carriers are equal, and the material as a whole has no charge.

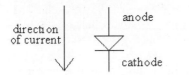

Junctions and diodes

A junction is an area inside a crystal where P-type material meets N-type material. You cannot create a junction by taking a piece of N-type semiconductor and placing it in contact with a piece of P-type material. The crystal has to be created so that these opposite types meet within the crystal, forming a junction. If we connect a wire to each part of the crystal we shall have the component we call a *diode*. The symbol for a diode is illustrated here, left.

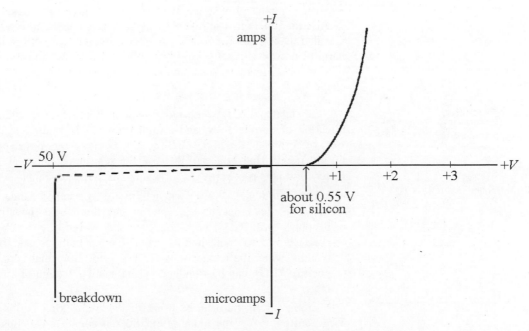

Figure 8.1 A typical characteristic for a diode using P- and N-type silicon (a silicon diode). Note the different scales used for voltage

A diode does not obey Ohms law, so that we cannot use $V = RI$ or any other simple formula to find a relationship between current and voltage. The relationship must be read from a graph, and a graph of this type is called a characteristic. A typical characteristic for a semiconductor diode is illustrated in Figure 8.1, assuming a diode made from P-type and N-type silicon. This shows that a diode is a one-way path for current.

The characteristic is really made up from two separate graphs. The right-hand side is the graph for current flow, meaning that the diode has its P-type material connected to the positive side of a supply. This is referred to as the *forward* direction, and the external voltage is called a *forward bias*.

The graph axes are scaled in terms of small voltages and currents, and the graph shows no current flowing until the forward voltage is about 0.6 V. When the diode starts to conduct, the graph shape is not the straight line that you would expect of a resistor, but a curve (an *exponential* curve). The curvature is upwards, so that the current value is multiplied for each small increase in voltage. There is no constant value of resistance, the behaviour is as if the resistance decreased as the voltage was increased.

When the voltage is applied in the reverse direction, so as to make the junction non-conducting, the voltage scale on the graph has to be changed in order to show anything happening. For this particular diode, nothing is measurable until the voltage reaches a high value, around 100 V in this example, at which point, the diode becomes conducting with a very low resistance. Unless there is some means of limiting the current flow, such as a resistor connected in series, the current that passes in this condition will destroy the junction. The reverse voltage that is needed to cause this effect is called the *breakdown voltage*. Applying breakdown voltage to a diode is harmful only if excessive current is allowed to flow, and the principle of reverse breakdown is used deliberately in Zener diodes, see later, this chapter.

Testing a diode

We can test a diode by checking that it has a low resistance for one direction of current flow, and a high (unreadably high) resistance to the flow of current in the opposite direction. Some meters have special diode-testing ranges, but you can also use the resistance range of an ordinary meter for this purpose.

- When some meters are switched to the resistance range, the red (positive) lead becomes negative, and the black (negative) lead becomes positive. The diode conducts when the black lead is connected to the anode and the red lead to the cathode, and there is no reading when the connections are reversed. Note that some diodes, notably LEDs, can be damaged by the battery voltage of a resistance meter.

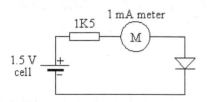

A simple circuit that can be used for diode checking is illustrated here. When current flows, the anode of the diode is connected as illustrated.

Diodes can fail open-circuit (o/c) or short-circuit (s/c) and either type of failure is easily detected by testing with a multimeter. An o/c diode will pass no current in either direction; it has a high resistance in both directions. An s/c diode will pass current equally easily in both directions.

Activity 8.1

Use an ohmmeter to identify the anode and the cathode connections of an un-marked diode. A good diode should indicate a very high resistance in the reverse direction, and a low resistance in the forward direction. Remember that many multimeters operate with reverse polarity when the resistance range is selected, and be careful to avoid touching the leads when using the highest resistance ranges.

Activity 8.2

Connect the circuit illustrated here, using a germanium diode. Turn the potentiometer control so that the voltage across the diode will be zero when the circuit is switched on. Make sure that you know the scales of current and voltage you are using – a 1 V or 1.5 V scale of voltage and a 10 mA scale of current. The voltmeter should be of the high-resistance type.

Watching the meter scales, slowly increase the voltage. Note the reading on the voltmeter when the first trace of current flow is detected. This value of voltage is called the *contact potential*. Below this value, the diode does not conduct.

Note the voltage readings for currents of 1 mA, 5 mA, and 10 mA. Draw a graph and ask yourself if the diode obeys Ohm's Law with a constant resistance value? Now repeat the readings using a silicon diode. What differences do you notice? Note that this circuit measures the voltage across the current meter, but this should be very small compared to the voltage across the diode.

Question 8.1

When does a diode have its highest forward resistance, at high current or at low current?

Diode types

Diodes of many types exist, and we can classify them roughly as signal, power, Zener, LED, and photo types.

Signal diodes are used for tasks such as demodulation (see Chapter 16), using a high-frequency signal into the diode. Signal diodes need not have a very low resistance when conducting, and it is more important to have very low values of stray capacitance between anode and cathode.

Rectifier diodes are used in power supply units (PSUs), see later, this chapter. The principle is like commutation, changing over the connections to an a.c. supply twice on each cycle so that only the positive peaks are passed. Because the cathode of the rectifier diode is the positive d.c. output terminal it is often marked with a + sign, or coloured red.

Schottky diodes use a different type of junction formed between aluminium and just one type of silicon. This has the advantage of conducting at a low forward voltage, and is used in modern power supplies, particularly for computers, to reduce heat dissipation.

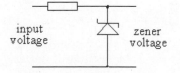

Zener diodes are used in quite a different way. They are connected in the reverse direction, with the cathode positive. A Zener diode has a low breakdown voltage, so that it must be used with a resistor in series to prevent damage. In this type of circuit, illustrated left, the voltage across the diode is almost perfectly constant, even if the current varies considerably. This type of diode is used to obtain a steady voltage for regulators, see later. In the illustration, the input is a d.c. voltage higher than the Zener voltage, the output is the Zener voltage, which does not change if the input voltage changes.

LEDs are diodes that emit light when the diode conducts in the normal forward direction. The usual colours are red or green, and some diodes can give a yellow light. There are also LEDs whose light changes colour when the voltage is changed. All LEDs have a *very low* breakdown voltage, so that even connecting them to a 1.5 V cell in the opposite direction can cause irreparable damage.

- Note that you can obtain information on diode connections and ratings from manufacturers data-books, from independently published data-books, or from the Internet, using a search engine.

Activity 8.3

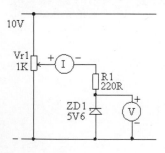

Connect a Zener diode in the circuit illustrated here, making sure that the cathode of the diode is connected towards the positive pole of the supply. The voltmeter should be set to the 10 V range, and the milliammeter to the 10 mA range. Switch on, and adjust the potentiometer so that the voltage across the diode can be read for currents of 1 mA, 5 mA and 10 mA. Note the voltage readings for each of these current values.

If the first reading had been taken across a resistor, what would the reading for 10 mA have been? (HINT – find the resistance value and then use the potential-divider formula.)

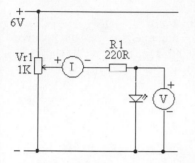

Activity 8.4

Connect an LED in the circuit illustrated here. The voltmeter should be set to the 5 V range and the milliammeter to the 10 mA range. Be very careful to connect the LED with the correct polarity.

Starting with the potentiometer at its lowest setting, switch on and slowly turn up the voltage until the diode conducts. Note the voltage across the diode when the current first starts flowing. Examine the light output at current values of 1 mA, 5 mA and 10 mA and note the value of forward voltage at these currents.

Question 8.2

Why would a rectifier diode be unsuitable for use with high-frequency signals?

Photo diodes are used with reverse bias, but they will conduct when they are struck by light. The currents that can pass are very small, a few microamps, so that amplification is needed. Photo diodes are used in light detectors when a fast response is needed – more sensitive devices are used if a slow response is acceptable.

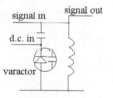

Varicap or varactor diodes are used for tuning high-frequency circuits. These diodes are used with reverse bias, and are designed so that the capacitance between anode and cathode changes as the amount of bias is changed. The varactor is used along with a fixed capacitor as part of a tuned circuit, so that altering the steady bias voltage will alter the tuned frequency.

Assorted diode circuits

Clipping, d.c. restoring and limiting are processes that can be carried out on waveforms of any shape, but which need diodes as well as the resistors, capacitors and inductors that are used in wave filters.

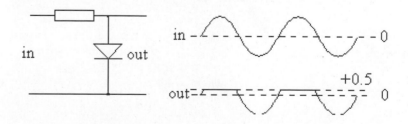

Figure 8.2   A simple clipping circuit

Clipping means the removal of part of either peak of a wave. A simple clipping circuit is shown in Figure 8.2. The silicon diode does not conduct when its anode is negative, and will conduct only when its anode voltage reaches about 0.5 V positive. The clipping circuit as a whole acts like a potential divider, with the diode resistance being very high for voltages of less than 0.5 V. The waveform is therefore clipped at about +0.5 V.

Such a clipped waveform will have a d.c. component unless both positive and negative peaks are clipped equally. A circuit for doing this is shown in Figure 8.3.

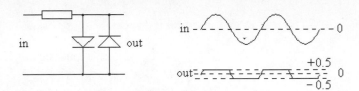

Figure 8.3   Two-way clipping

Restoration (or d.c. restoration) is the process of restoring to a signal the d.c. level which it will have lost if it has been passed through a capacitor or a transformer. Figure 8.4 shows a typical application. The unidirectional waveform has been passed through a capacitor, so losing its d.c. component. A simple d.c. restoring circuit then replaces the missing d.c. and the d.c. level is again present.

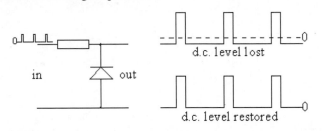

Figure 8.4   How a d.c. restoration circuit works

The diode in the circuit does not conduct when the waveform is positive, but passes current when the wave swings negative relative to the zero (earth) line. When the diode conducts, capacitor C is charged positively and has a positive voltage between its plates when the diode ceases to conduct. This positive voltage is equal to the missing d.c. voltage, and the diode will thereafter conduct only as required on the negative peaks to keep the voltage across the capacitor at the correct level.

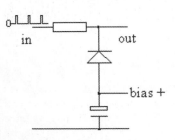

A d.c. restoration circuit which ensures that the minimum voltage of a waveform is higher than earth (zero voltage) is shown, left. This raises the minimum level by V volts, the bias battery voltage.

Limiting means restricting the peak amplitude of a signal, preferably without clipping it to do so. It can be carried out with the aid of a voltage divider, one part of which is a controllable resistance (which is often either a transistor or a FET). A feature of both these types of semiconductor is that the resistance between two of their three electrodes depends on the voltage at the third electrode.

A typical limiting circuit is shown in block form in Figure 8.5. The amplitude of the waveform is used to generate a d.c. voltage. This d.c. voltage is used to control the resistance of a transistor or FET used as a controllable resistor in a voltage divider. If the amplitude of the signal increases, the d.c. level increases also and this increased d.c. voltage is used to decrease the resistance of the controllable resistor, so reducing the amplitude of the signal leaving the divider. The effect is that, whatever the amplitude of the input signal, the output amplitude becomes almost constant.

The circuit can be adjusted so that only signals above a preset voltage limit, the *clipping level*, are treated in this way. Limiting circuits of this type are widely used in tape recorders as automatic recording level controls or as overload prevention circuits, and in radio transmitters to prevent over-modulation.

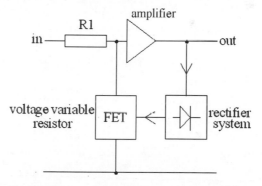

Figure 8.5   A limiting amplifier in block diagram form

Question 8.3

Why are limiting amplifiers often used in tape recording and radio transmitter applications?

Mains supplies

The cheapest method of operating any electronic circuit whose power consumption is more than a few milliwatts is by means of a power supply unit (*PSU*) that draws its energy from the a.c. mains supply. The purpose of a PSU is to raise or, more usually, lower the mains supply voltage to an amplitude suitable for the application, and then to convert this a.c. voltage into a steady d.c. output. Sometimes additional circuits called regulators or stabilizers are needed to ensure that the d.c. output voltage remains steady even when mains voltage varies, or if the amount of current taken from the PSU is altered.

The a.c. voltage conversion is carried out by a *transformer*. A.c. voltage from the mains is applied to the primary of the transformer, and induces a voltage between the terminals of the transformer secondary. The value of this secondary voltage depends on the relative number of turns in the two windings. Most modern transistor equipment requires a lower secondary voltage, usually within the range 6 V to 50 V.

The first stage in the conversion of the low a.c. voltage into d.c. is carried out by a diode rectifier, or series of rectifiers, which alter the waveshape of the a.c. into one of the waveshapes shown in Figure 8.6 of which the full-wave (bridge) rectifier arrangement is preferred for most applications.

The result of rectification is to produce the waveforms shown on the right-hand side of Figure 8.6 which are of a voltage in one polarity only. This voltage will have an average value which is equivalent to d.c. of that same voltage, but is not a smooth steady d.c. voltage.

To convert this output into smooth d.c., a reservoir capacitor and filter must be used. The reservoir capacitor (C_1) is charged by the current from the rectifier(s), and it then supplies current to the circuit every time the voltage output from the rectifiers drops. In this the waveform is smoothed into d.c. plus a small amount of remaining a.c. called *ripple*. Another capacitor (shown as C_2) can be used for further smoothing.

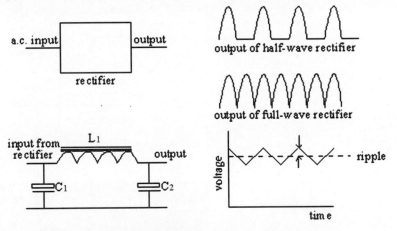

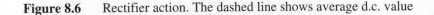

Figure 8.6   Rectifier action. The dashed line shows average d.c. value

This ripple is almost zero when only a little current is taken from the PSU, but it increases considerably when more current is taken. If it becomes unacceptably high, a larger value of reservoir capacitor must be used, or additional filter stages must be added. Another way to reduce ripple is to add a *regulator* or *stabilizer* circuit, see later.

Half-wave and full-wave rectification comparison

When no load current is taken from a rectifier circuit, the output from a half-wave circuit, as measured by a d.c. voltmeter, will be identical to that from a full-wave circuit supplied with the same d.c. voltage. When a load current flows, however, the output voltage of a half-wave rectifier circuit will drop by a greater amount than will that of the full-wave rectifier circuit, unless a much larger value of reservoir capacitor is used. Half-wave rectifier circuits are used in microwave cookers, but are seldom used in other equipment.

The a.c. ripple which appears when the load current is taken from a rectifier circuit will be at mains frequency when the rectifier is a half-wave one, but at twice mains frequency when the rectifier is a full-wave circuit When the load current taken is very large, the voltage output from a half-wave supply will fall to about half the value of that from a full-wave supply connected to the same a.c. voltage. The most common type of rectifier circuit used now is the full-wave bridge type, which is usually obtained in a package form with two (a.c.) inputs and two (d.c.) outputs. The drawings, left, illustrate typical bridge rectifier packages.

Neither type of supply has good regulation, though they are adequate for many purposes. *Regulation* is the quantity that measures the drop of output voltage which occurs when load current is taken from the supply circuit. A perfectly regulated supply would have constant output voltage whatever the noted load current; in other words, it would have a zero value of internal resistance.

Failure in any part of a conventional PSU is easy to diagnose. Symptoms are complete loss of output voltage, a drop in output voltage, excessive ripple or unusually poor regulation. Complete failure can be caused by a blown fuse or by an o/c transformer winding. A short-circuit in a half-wave rectifier will also bring output to zero. A short circuit winding in the transformer will, in addition, generate excessive heat.

A drop in voltage, usually accompanied by excessive ripple can be caused by the failure of one of the rectifiers in a full-wave set, or by an o/c reservoir capacitor. Poor regulation can also be caused by an o/c reservoir capacitor, or by a diode developing unduly high resistance.

Activity 8.5

Connect the a.c. rectifier/reservoir circuits shown in Figure 8.7. (Note that these are intended for use with 12 V a.c. supplies only – they must *never* be connected to the mains.)

Use the a.c. voltage range of a multimeter to set the a.c. input to 12 V, and measure the output voltage of each circuit. Note down the readings under appropriate headings. Note also the output readings when the reservoir capacitors are temporarily disconnected. Compare the ripple amplitude and frequency using an oscilloscope.

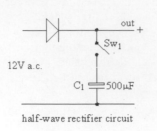

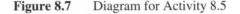

half-wave rectifier circuit

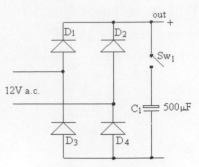

full-wave bridge rectifier circuit

Figure 8.7   Diagram for Activity 8.5

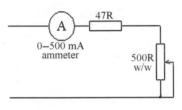

Activity 8.6

Now connect to each output in turn the load circuit illustrated left. Measure the output voltages at various load currents from 50 mA to 200 mA for both rectifier circuits, and draw up a table of output voltage and load current. Note the effect on the output voltage and waveforms when any one diode is open circuited (so reverting to a half-wave circuit). Draw a graph of output voltage plotted against load current for each type of rectifier circuit.

Question 8.4

A full-wave rectifier circuit fails, with the voltage under load dropping, but still delivering well-smoothed d.c. What is the most likely cause?

A third type of rectifier that was used to a considerable extent at one time is the bi-phase half-wave, see circuit, left. This provides the same output waveform as the full-wave bridge circuit, but uses two rectifier diodes only. The disadvantage is that a transformer with a centre-tapped secondary is needed, and because bridge rectifier packages are so cheap now, the bi-phase half-wave circuit is hardly ever used.

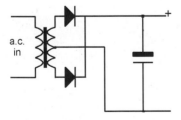

Regulators

To keep the output voltage of a rectifier circuit steady despite variations in the a.c. mains input or in the amount of current drawn at the output, it is sometimes possible to use a subsequent *regulator* or *stabilizer* circuit. This can only be done however, if the voltage output from the filter is several volts higher than the voltage output required from the regulator itself. A regulator **cannot** restore lost voltage. It can only control the voltage at its output; and it cannot operate at all if the voltage from the filter stage is lower than the voltage setting of the regulator (see Figure 8.8).

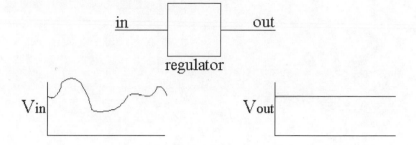

Figure 8.8   The action of a voltage regulator

A fixed voltage regulator circuit will produce its fixed output voltage only if the input voltage never falls below the minimum needed to operate the regulator. A variable output regulator is a circuit whose steady d.c. output is controlled by the setting of a potentiometer but which will remain stabilized at whatever voltage is set by that potentiometer. Again, regulation is possible only if the input d.c. voltage is high enough to allow the regulator to function correctly. If the input voltage is inadequately smoothed or is too low to permit correct regulator action, there will be no true regulation.

In the event of suspected regulator failure, the waveform and voltage of the input should be checked before any other steps are taken.

Regulator circuits are more liable to failure than are simple rectifier/reservoir supplies because of their additional active components. In the simple circuit shown, the failure would be indicated either by zero output or by an unregulated output at about the same voltage as the input.

Other types of circuits for regulating a.c. also exist. Saturable transformer regulators consist of specially wound transformers whose output can be controlled by a d.c. circuit. The d.c. is provided by rectifying part of the a.c. output so that the output becomes self regulating. Take care when you are either connecting or disconnecting such regulators, for they contain charged capacitors which can be at a high voltage. Regulators of this type are commonly used in photographic processing plants to regulate the supply to enlarger lamps so that colour processing becomes more consistent.

Thyristor or triac a.c. regulators use electronic switching components to regulate the average power of a.c. Note that all types of a.c. regulator distort the shape of the mains waveform, so that it is normal to connect filter circuits to both the input and output leads of these regulators to minimize the transmission of radio frequency interference (RFI).

Activity 8.7

Construct the regulator circuit shown in Figure 8.9, following. Tr_1 is of the type MJE3055, a power transistor, which is housed in a convenient flat-packed casing. Bolt the metal face of the transistor through the hole in the casing to a 6 cm. square sheet of 14 gauge aluminium or to a pre-shaped heatsink. The other two components, the resistor R_1 and the Zener diode ZD_1, can be mounted on tagstrip or matrix board.

Connect the regulator as shown in Figure 8.9 to the full-wave rectifier/reservoir circuit, and switch on. Measure the output voltage resulting from various current flows.

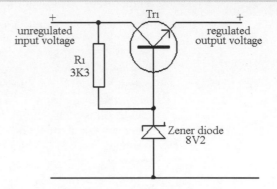

Figure 8.9   Diagram for Activity 8.7

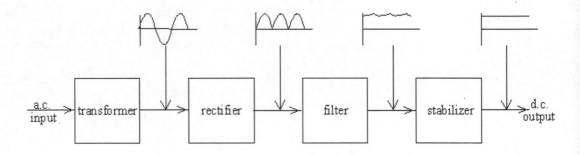

Figure 8.10   Block diagram of power supply

Modern equipment makes use of a different form of power supply, the switch-mode supply (SMS) circuit. This is very much more complex, and is difficult to service because a fault cannot easily be located. The principle is to rectify mains voltage and use this supply to operate a high-frequency oscillator, whose output is controlled by a feedback circuit. This signal at high frequency is then rectified to provide the d.c. output. The switch-mode power supply is particularly suited to providing low-voltage high-current supplies with excellent regulation, and is used in all computers. Switch mode circuits are also used extensively in TV receivers, and are likely to replace the older type of regulated supply for other applications also.

Activity 8.8

Construct the power supply system shown in block diagram form illustrated on the previous page in progressive stages, starting with a suitable resistive load connected across the mains transformer secondary winding, taking specific note of the correct polarity of electrolytic capacitors. Using a multirange meter and oscilloscope, study and sketch the voltage levels and waveforms obtained. Measuring the ripple amplitude and d.c. level. Repeat these steps as the rectifier, filter and IC regulator units are added.

Suitable component values are as follows:

Transformer, mains input, 12 V at 100 mA secondary winding.

Reservoir and smoothing capacitors, 500 µF, 50 V d.c. working.

Smoothing choke, 5 Henries.

Regulator, 12 V input, 8–10 V output at 100 mA.

Suitable load, 82R, 5 W resistor. If not available, a 15 watt 240 V mains lamp can be used as a substitute.

Draw a regulation curve for the complete regulated unit by plotting output voltage against load current.

Question 8.5

A regulator circuit appears to be working perfectly when tested, but when it is connected to a rectifier circuit and a load the output voltage is low. Can you think of a simple explanation?

Battery charging

One important use for rectifier circuits is battery charging, and two main types of circuits are used. Lead-acid cells need to be recharged from a constant-voltage supply, so that when the cell is fully charged, its voltage is the same as that of the charger, and no more current passes. By contrast, nickel-cadmium cells must be charged at constant current, with the current switched off when the cell voltage reaches its maximum. Constant-current charging is needed so that excessive current cannot pass when the cell voltage is low. Simple transformer–rectifier circuits can be used for constant-voltage charging, but some form of regulation is preferable, particularly for sealed lead-acid cells.

Nickel-metal-hydride cells need a more complicated charger circuit, and one typical method charges at around 10 per cent of the maximum rate, with the charging ended after a set time. Some types of cells include a temperature sensor that will open-circuit the cell when either charge or discharge currents cause excessive heating. For some applications trickle charging at 0.03 per cent of maximum can be used for an indefinite period.

Lithium-ion cells can be charged at a slow rate using trickle-chargers intended for other cell types, but for rapid charging they require a specialized charger that carries out a cycle of charging according to the manufacturer's instructions.

Note that there is no such thing as a completely universal battery charger and though several microprocessor-controlled chargers are available they cannot be used on all types of cells or on mixed sets of cells.

Answers to questions

8.1   At low current.

8.2   The capacitance of a rectifier diode is comparatively high, and its response is too slow for high-frequency signals.

8.3   Excessive input signals to a tape-recorder or a transmitter will cause severe distortion, so that the gain-reducing action of a limiting amplifier is a useful safeguard.

8.4   One diode is open circuit.

8.5   The output from the rectifier circuit is at too low a voltage for the regulator to operate correctly.

9 Transistors and ICs

Transistors
The bipolar transistor

Though you may seldom see transistors used as separate components in modern circuits, it is important to know how they work, because they form the basis of the integrated circuits (ICs) that are used in virtually all electronic circuits today.

The *bipolar junction transistor* (BJT) is a device that makes use of two junctions in a crystal with a very thin layer between the junctions. This thin layer is called the base, and the type of transistor depends on whether this base layer is made from P-type or from N-type material. If the base layer is of N-type material, the transistor is said to be a PNP type, and if the base layer is of P-type material, the transistor is said to be an NPN type. The differences lie in the polarity of power supplies and signals rather than in the way that the transistors act. For most of this chapter, we shall concentrate on the NPN type of transistor, simply because it is more widely used. The drawing shows the symbols for the NPN transistor (a) and its opposite, the PNP type (b).

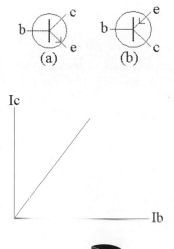

The transistor has three connections, called emitter, base and collector. For the NPN type, the transistor is connected so that the collector is positive, with the emitter at a lower voltage, often earth (zero) voltage. With no connection to the base, this arrangement does not pass current. When a small positive voltage, about 0.6 V for a silicon transistor, is applied to the base, a base to emitter current, Ib, will flow, and when a base current flows, a current, Ic, also flows between the collector and the emitter. The drawing, left, shows these quantities graphed.

This *collector current*, Ic, is always proportional to the base current; a typical value is that the current between emitter and collector is some 300 times the current between base and emitter. The exact size of this ratio depends on how thin the base layer is, and it varies considerably from one transistor to another, even when the transistors are mass-produced in the same batch. This value is variously known as forward current gain or h_{fe}.

The action of the PNP transistor is similar, except that polarities are changed, with the collector negative, and a negative bias voltage applied to the base.

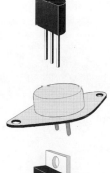

Transistor types include small-signal transistors, RF transistors, power transistors and switching transistors. The small-signal types are used for amplifiers that work at frequencies ranging from audio to the lower RF. The RF transistors are intended for the higher radio frequencies, and some specialized types can be used at SHF (in satellite receiving dishes, for example). Power transistors are used for amplifier outputs, for driving solenoids and motors, for TV use, and for transmitter outputs. Switching transistors are used with digital circuit to switch large amounts of current or high voltages in very short times of the order of nanoseconds.

Transistors exist in a huge variety of sizes and shapes, and it is important to be able to identify the electrodes (emitter, base and collector). Data on transistors can be obtained from manufacturers' books, independent source-books, or Internet sources. Two well-known source-books are *Tower's International Transistor Selector* and the *International Transistor Equivalents Guide* (by Michaels).

If no data is obtainable, you can use a resistance meter to identify the connection, because for an NPN transistor the base will conduct to both other electrodes when the base is positive. If you connect the positive lead of the resistance meter to one lead, and test by holding the negative lead to the other two in turn, the positive lead is on the base if both tests show a low resistance. You may have to try three times until you clip the positive lead to the correct connection.

Transistor types are coded either by the US MIL specification or by the European (Pro-Electron) lettering system. The US system starts with 2N followed by a code number; the 2N portion identifies the device as a transistor (1N indicates a diode). The full description can be found only if you can identify the manufacturer by looking up the code number.

The European system, used for all types of semiconductors, uses two or three letters followed by a number. The letters indicate the type of device, with the numbers indicating how recent the design is. The table below shows the meanings of the letters.

Table 9.1   European coding letters

The first letter indicates the type of semiconductor material used.

A	detector diode, high speed diode, mixer diode	B	variable capacitance diode
C	transistor for a.f., not power amplification	D	power transistor for AF
E	tunnel diode	F	transistor for RF amplification (not power)
G	multiple transistor with dissimilar devices on same chip	L	power RF transistor.
P	photosensitive device or other detector of radiated energy	Q	radiation emitting device such as light-emitting diode.
R	thyristor or other switching device with specified breakdown characteristic. Not high power	S	switching transistor, not power types
T	thyristor or other controller, high power	U	power transistor for switching
X	multiplier diode such as varactor or step recovery diode	Y	rectifier diode, booster diode, efficiency diode
Z	voltage reference or voltage regulator (Zener) diode		

The second letter indicates the construction or use of the device.

A   Germanium junction

B   Silicon junction

C   Gallium Arsenide or similar compounds forming junction

D   Indium Antimonide or similar compounds forming junction

R   Photoconductor, or other device with no junction

The figures or letters following indicate the design; A three figure serial number is used for 'consumer types', used in domestic radio, TV, tape recorders, audio etc. A serial consisting of a letter (Z Y, X, W, etc.) followed by two figures means a device for professional use (transmission etc.). The European system is much more informative, because you can tell the type of device immediately from the code. The 1N, 2N system tells you only whether the device is a diode or a transistor.

Activity 9.1

Test both PNP and NPN transistors using a multimeter. For an NPN transistor you should find a low resistance when the meter is connected with base positive and emitter or collector negative. For the PNP transistor, the low resistance reading will occur for base negative and emitter or collector positive. All other readings should show high resistance. Make also a set of readings on transistors that are known to be faulty with either s/c or o/c connections.

Heatsinks

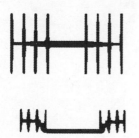

When a transistor conducts, the current that flows between the collector and the emitter causes heat to be dissipated. Most of this heat comes from the collector-to-base junction, and because the heat is concentrated in a small area, it can cause damage, eventually melting the junction and destroying the transistor instantly.

When transistors are used for circuits in which large amounts of current flow, some way of removing this heat is needed, and is provided by using a *heatsink*. A heatsink, illustrated left, is a metal plate, often quite heavy and fitted with fins. The transistor is bolted to this plate, usually with a thin layer of heatsink grease, and possibly a mica insulating washer, between the transistor and the plate. This grease is a silicone grease that conducts heat well, and is also a good electrical insulator. The heat from the junction spreads to the plate then so to the air, preventing the transistor from becoming too hot. The use of a heatsink allows more power to be dissipated, but the power must be kept within the limits that the heatsink has been designed for, otherwise damage can still be done. Heatsinks for computer ICs often incorporate fans, as noted earlier.

- Note that modern power transistors can run so hot that they will cause burns if you touch them. A fuse cannot protect a transistor. A fuse takes several milliseconds to blow, but a transistor can have its junctions vaporized in less than a microsecond.

Simple BJT amplifier

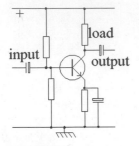

The FET

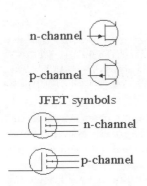

n-channel

p-channel

JFET symbols

n-channel

p-channel

MOSFET symbols

MOSFET = Metal-Oxide-Semiconductor FET

The illustration shows a simple voltage amplifier constructed using a single bipolar junction transistor, in this example, an NPN type. The input signal is taken to the base, whose steady voltage level is fixed by two resistors. The resistor in the emitter circuit controls the steady (bias) current. The load resistor is placed in the collector circuit, and the output signal is taken from the collector. The same principles are used when the load is an inductor or a tuned circuit, or when the input is taken from a transformer.

Though this form of simple amplifier is seldom seen now because of the extensive use of ICs, you should be able to identify the input and output points, and the load.

The *field-effect transistor*, or *FET*, is as old a device as the bipolar transistor, dating from 1948, but it was not commercially exploited until much later. The principle of any type of FET is to allow current to flow between two terminals, the source and the drain, which are connected by a thin strip of doped semiconductor called the channel. All of the semiconductor material between the source and the drain terminals will be of the same doping, which can be either P-type or N-type, though N-type is more common for the channel. The choice of doping determines the type of carrier that will be used, so by selecting N-type doping, the carriers will be electrons.

The original type was the junction FET, but these are now uncommon. The MOSFET uses a channel, but the gate connection is to a strip of conductor that forms one plate of a capacitor, with the channel forming the other plate. The gate is therefore insulated from the channel, and no current flows between the gate and the other electrodes. You should never try to use a resistance meter on a MOSFET. Typical symbols are illustrated left. Though power MOSFETs are used in hi-fi amplifiers, you are not likely to encounter MOSFETs other than as ICs using this type of construction.

Question 9.1

Why is it dangerous to measure the gate to source resistance of a FET with a standard ohms meter?

Integrated circuits (ICs)
Integration

A transistor of any type, bipolar or FET, is manufactured by a set of operations on a thin slice, or wafer, of silicon crystal. These operations include selective doping and oxidation (for insulation), and the areas that are affected can be controlled by the use of metal masks placed over the semiconductor. By using different amounts of doping of a strip of semiconductor material, the resistance of the strip can be controlled, so that

it is possible to create a resistor on a semiconductor chip. In addition, because silicon oxide is an excellent insulator, it is possible to make capacitors by doping the semiconductor to make a connection, oxidizing to create an insulating layer, and depositing metal or semiconductor over the insulation to form another connection.

Since transistors, resistors and capacitors can be formed on a silicon crystal chip, then, and connections between these components can also be formed, it is possible to manufacture complete circuits on the surface of a silicon chip. The transistor types can be bipolar or FETs, or a mixture of both, but FETs are preferred because they are easier to fabricate in very small sizes.

The advantages of forming a complete circuit, as compared to the discrete circuit in which separate components are connected to a printed circuit board (PCB), are very great. The obvious advantages are that very small complete circuits can be made. It is possible to pack huge numbers of components on to one chip simply by making the components very small, and the technology of creating very small components has advanced spectacularly. The idea of packing three million FET transistors on to one chip might have seemed unbelievable only a few years ago (and it's still hard to imagine even now), but this is the extent that we have come to. The advantage that this provides is that we can manufacture very complicated circuits that simply would not be economic to make in discrete form.

Another advantage of integrated construction is of cost. The cost of making an integrated circuit that contains 50 000 transistors is, after the tooling has been paid for, much the same as the cost of making a single transistor. This is the basis of the £2.50 calculator, the TV remote control, and the small computer. The cost of integrated circuits has been steadily reducing with each improvement in the technology, so that prices of computers go down even faster than the prices of cars go up. The greatest advantage from the use of integration, however, has been reliability, and it was the demand for 100 per cent reliability in the electronics for the space missions that fired the demand for ICs.

Modern ICs can be formed with several million transistors on one chip, and with very elaborate connections. Such ICs are used in computing, communications, and for the decoders for digital radio and TV.

Consider a conventional circuit that uses 20 transistors, some 50 resistors and a few other components. Each transistor has three terminals, each resistor has two terminals, so that the circuit contains some 160 soldered connections. One faulty connection, one faulty component, will create a fault condition, and the greater the number of components and connections there are the more likely it is that a fault will develop.

By contrast, any IC which will do the same work might have only four terminals, so that it is connected to a PCB at four points only. The IC is a single component which can be tested – if it works satisfactorily it will be used. Its reliability should be at least as good as that of a single transistor, with the bonus that each internal part will always be working under ideal conditions if the IC has been correctly designed.

The reliability of a single component is 20 times better than the reliability of 20 components that depend on each other. If the IC does the work of a thousand components, its reliability will be about a thousand times better than that of a single component and so on. This enormous improvement in reliability, many times greater than any other advance in reliability ever made, is the main reason for the overwhelming use of ICs. When electronic equipment fails, the first thing to check is the connections to the mains plug, not the state of the ICs.

For these reasons, circuits that use separate (discrete) transistors are becoming a rarity in modern electronics, which is why so comparatively few circuit examples have been shown using transistors. When we work with ICs, circuits become very different. We may still be concerned with d.c. bias voltages, but usually at just one terminal. We are still concerned with signal amplitudes and waveshapes, but only at the input and the output of the IC. What goes on inside the IC is a matter for the manufacturer, and we simply use it as advised.

The IC is a complete circuit, and servicing amounts to deciding whether the IC is correctly dealing with the signal that is applied to its input. If it does not, then it has to be replaced. If the input and output signals are correct, the fault lies elsewhere. The use of ICs has not complicated servicing, nor has it made designers redundant. On the contrary, like all technical advances, it has resulted in a huge increase in the amount of items to service and the variety of goods that can be designed.

IC examples

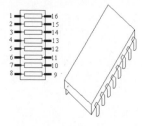

One type of integrated component contains no transistors, but consists entirely of resistors. The aim is to provide a single component for a circuit that requires a large number of identical resistors, and these integrated resistors can be packaged as SIL (single in line) or DIL (dual in line), referring to the line(s) of pins that are used for connections. Typically, these units will contain anything between 5 and 15 resistors; one very common value is eight. These resistor units are often called *thick-film* circuits because of their construction from comparatively thick metal films.

Question 9.2

Why are resistors sometimes constructed as thick-film devices? What other components can be constructed using thick-film techniques?

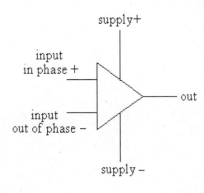

For most purposes an IC of the type called the operational amplifier (usually shortened to *op-amp*) will be used in place of a multi-stage amplifier constructed from individual transistors. You do not usually know the internal circuit of such an IC, and the important point to recognize is that the gain and other features of the op-amp will be set by external components such as resistors. The voltage gain of the circuit alone is typically very high, 100 000 or more. The drawing shows a typical op-amp symbol.

The point about performance being controlled by resistors and other external components is true also of most amplifiers that have been constructed by using individual transistors, but since the op-amp consists of one single component, it is very much easier to see which components control its action, since only these components are visible on the circuit board.

Operational amplifiers can be packaged as a single unit, or as a set. Single units are usually packaged as 8-pin DIL, but the multiple unit will be placed in larger packages.

Voltage regulators are another type of IC that has totally replaced the older discrete units. The voltage regulator IC is usually contained in a package with a metal tab that can be connected to a heatsink, and the regulator has three connections. One connection is earthed, one is used for the unregulated voltage input, and the third provides the regulated output. Regulator ICs are made for fixed voltages, typically 5 V, 9 V and 15 V.

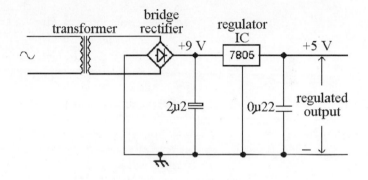

Figure 9.1   A voltage regulator circuit using an IC

In addition to multi-purpose ICs, manufacturers use application-specific ICs (ASICs). These are designed and built to order for some specific part of a hifi amplifier, TV receiver, computer, etc. Some of these use large packages, sometimes of the square type that is favoured for microprocessors. Such circuits are also likely to use a large number of pins.

The largest class of ICs is the digital type, including the microprocessor and memory chips that are produced in such enormous numbers. These types of ICs were in use before ICs started to be available for other purposes, so that the digital types are in many ways more advanced in design and construction. They are almost all of FET construction, and some use 420 or more pins in the package. The thermal dissipation of computer microprocessor chips is such that a fan and heatsink assembly must be clamped over the chip.

Packaging, pin numbers and coding

ICs are much too small to be used directly, so that packages are used. The tiny IC chip is embedded in plastic, and metal wires connect it to pins that are rugged enough for insertion into printed circuits. The most common packaging system is the DIL type, with two rows of pins. On this type of

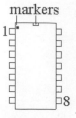

markers

1

8

packaging, the pins are numbered starting from Pin 1, which is located by a punch-mark, or from its position, and the numbering continues down one side and up the other side. SMC ICs are designed for *surface mounting*, and are very small, often no larger than a blob of solder. Note that pin numbering is always shown as seen looking down on the mounted IC, not from the pin side.

Data on ICs can be obtained from manufacturers' booklets, on computer data disks, or by way of other manuals. Coding of ICs will be by manufacturers' type-lettering and numbering, by Pro-electron coding letters and numbers, or for digital ICs following the 7400 or 4000 family series that is noted in Unit 5.

Handling of MOSFETS

MOSFETs and ICs that are based on MOSFETs require careful handling because of the risk of electrostatic damage. Electrostatic voltages arise when materials are rubbed together, and synthetic fibres such as nylon can rub on other materials to develop voltages of several kV. These voltages can cause sparks, and though the currents that flow are usually very small, the gate insulation of a MOSFET will not survive having such a voltage between gate and channel connections.

Though most devices incorporate over-voltage diodes that will conduct at a lower voltage than would damage the gate connections, there is always some risk of damage when devices are being inserted into boards. MOS devices that are soldered to boards are almost immune from accidental damage from electrostatic effects, because most boards will have resistors wired between the gate and the source or emitter terminals. Even a resistor of several thousands of MΩ is enough to reduce electrostatic voltages to a fraction of a volt because of the current that passes through the resistor.

Manufacturers of MOSFET devices need to use precautions such as using earthed metal benches, with staff wearing metal wrist clamps that are also earthed, but such precautions are seldom necessary for servicing purposes. The main risk is in opening the wrapping and taking out the IC, and you should try to avoid touching the pins. Sensitive ICs are often sold with the pins embedded in conducting plastic foam, and one useful precaution is to short the pins together (by using a metal clip or wrapping soft wire around them) until the device is soldered into place. If special precautions are specified for a device, these should be used.

Answers to questions

9.1   Because the test voltage may destroy the gate-to-channel insulation.

9.2   The components' values can be more easily matched to circuit needs. Blocks of identical small-value capacitors can also be constructed in this way.

Unit 4

Outcomes

1. Demonstrate an understanding of waves and waveforms and apply this knowledge safely in a practical situation.
2. Demonstrate an understanding of input and output transducers and apply this knowledge safely in a practical situation.
3. Demonstrate an understanding of electronic modules and apply this knowledge safely in a practical situation.

10 Waves and waveforms

Block diagrams

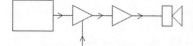

A *block diagram* of an electronic circuit is used to show what the circuit *does* rather than details of how it does it. A block diagram simplifies fault diagnosis, because it is often easier to diagnose, from a block diagram the general area of the circuit that might be faulty. This type of diagnosis can be much more difficult if you use a full circuit diagram. In addition, many systems that use ICs can be illustrated only by using block diagrams.

Another reason for the use of a block diagram is that the full circuit diagrams of some electronic devices, such as colour TV receivers and particularly computers, are far too large and complex to draw on a single sheet of paper. The circuit is divided into blocks, with one sheet of paper showing the full circuit diagram of one block, with all (or most) of the components and connections drawn in. The relation of each of these blocks to one another is then indicated on a full block diagram of the whole device.

Block diagrams can also be used to show the waveforms that ought to be present at various points in a circuit. These waveforms can be made visible with the aid of a cathode ray oscilloscope (CRO). Any significant change in the shape or the amplitude of a waveform can indicate a fault in that particular part of the circuit.

Both block diagrams and waveforms showing typical signals will be used throughout this book. Their purpose and use should be thoroughly understood at this stage.

Rectangular waves

You should at this point revise the section on waveforms in Chapter 5. In addition to the usual measurements of amplitude and frequency, rectangular waves, shown in Figure 10.1, require an additional measurement, called their *mark-to-space ratio*. It is possible for two rectangular waves that have the same period (and therefore the same frequency) and the same peak voltage levels to have quite different shapes and also quite different values of average voltage. For a rectangular wave, the mark is the length of time for which the voltage is positive and the space is the length of time for which the wave is either at zero or at a negative voltage.

A wave with a large mark-to-space ratio will thus have an average value nearly equal to its positive peak value, while a wave with a small mark-to-space ratio will have an average value nearly equal to its negative peak value (in the example shown in Figure 10.1, equal to almost zero). Varying the mark-to-space ratio of a rectangular wave of voltage is often used as a method of d.c. motor speed control because, using electronic methods, this is simpler than varying the amplitude and less likely to allow the motor to stall.

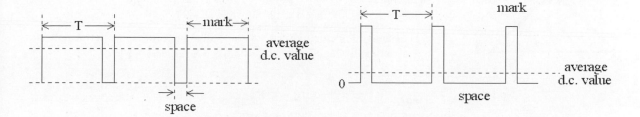

Figure 10.1   Rectangular waveforms or pulses

Square wave faults

Other measurements on a square wave that can be made with the help of an oscilloscope are its *rise* and *fall* times, its *sag* and its *overshoot* voltages. All of these are illustrated in Figure 10.2.

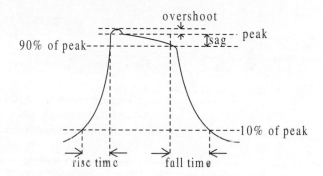

Figure 10.2   Imperfections in a rectangular pulse

The rise and fall times are, respectively, the times taken for the voltage to rise (or fall) between its 10 per cent and its 90 per cent voltage levels. Overshoot is the minor initial jump of voltage by which the leading edge of the square wave often exceeds the nominal peak level of the wave; while sag is the extent to which the peak voltage drops below nominal peak level before the trailing edge of the waveform arrives.

Wavelength

The *wavelength* of a wave is a quantity which applies to a wave that is travelling in space or along a wire. It can be measured easily only when the wavelength is short and is defined as the closest distance between neighbouring peaks of the wave. The measurement is illustrated in Figure 10.3.

In free space, the speed of all radio waves is identical at 3×10^8 metres per second. (10^8 is the figure 1 followed by eight zeros, so 3×10^8 is 300 million). This is the same speed as that of light, because light is also an electromagnetic wave. By comparison, the speed of sound in still air is about 332 m/s, but this is affected by movement of the air or by movement

of the source of the sound. We can write the relationship using the triangle as a reminder of the three possible forms of the equation

$$c = \lambda f \qquad \lambda = c/f \qquad f = c/\lambda$$

Wavelength calculations are of particular importance in the design and use of aerials for high frequency (mainly TV) signals.

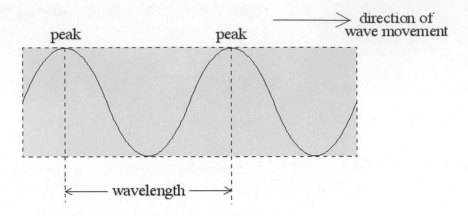

Figure 10.3   The wavelength of a transmitted wave

$\lambda = \dfrac{c}{f}$

The peak points of the wave are called *antinodes*. They can sometimes be located on wires with the aid of lamp indicators (see Figure 10.4); but the wavelength of a wave is usually a quantity that we calculate. The formula used is shown here, left, where λ (Greek letter lambda) is the wavelength in metres, c is the speed of wave travel in metres per second and f is the frequency of wave in Hz.

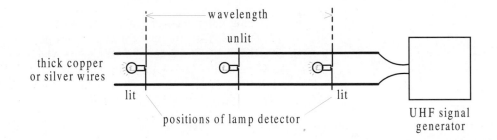

Figure 10.4   Measuring the wavelength of UHF waves travelling along wires

Example: What are the wavelengths, in free space, of waves of the following frequencies: (a) 1 MHz? (b) 30 MHz? (c) 900 MHz?

Solution: Use the equation $\lambda = c/f$ with c = 3×10^8 m/s. Then

(a) 1 MHz is 10^6 Hz. So $\lambda = 3 \times 10^8/10^6 = 3 \times 10^2 = 300$ metres.

(b) 30 MHz is 3×10^7 Hz. So $\lambda = 3 \times 10^8/3 \times 10^7 = 10$ metres.

(c) 900 MHz is 9×10^8 Hz. So $\lambda = 3 \times 10^8/9 \times 10^8 =$ one third of a metre, or 33.3 cm.

Question 10.1

A UHF transmission uses a frequency of 750 MHz. What is its wavelength?

A wave may be either pure a.c. or unidirectional. A pure a.c. wave has an *average* value of zero, so it can cause no deflection of a d.c. meter. A wave that is fed through a capacitor or taken from the secondary of a transformer (see Figure 10.5) is always of this type apart from certain short-lived (transient) effects which occur immediately after the circuit is switched on.

A *unidirectional* wave is one that has a steady average value that can be measured by a d.c. meter. Such a wave has a d.c. component, which means that its amplitude must be measured by a d.c. meter. The a.c. component of the wave can also be measured when the d.c. component is absent, by inserting a capacitor (which may have to be of large value) between the circuit and the a.c. meter. This assumes that a suitable a.c. meter is obtainable. Most a.c. meters will give correct readings only for sine wave waveforms.

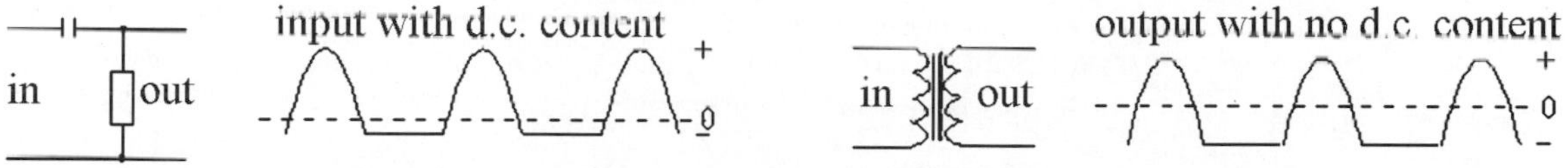

Figure 10.5   Circuits which transmit no d.c. component

Harmonics

A perfect sine wave has a single value of frequency, but waves that are not of sine wave shape (*non-sinusoidal waves*) contain other frequencies, called *harmonics*. An instrument called a *spectrum analyser* can detect what harmonic frequencies are present in such waves.

One frequency, called the *fundamental frequency*, is always present. It is the frequency equal to 1/period for the waveform. No frequencies of less than this fundamental value can exist for a wave, but higher frequencies which are whole number multiples of the fundamental frequency (i.e., twice, three times its value and so on) are often present. A pure sine wave consists of a fundamental only.

Adding time related waveforms

It is also possible to add together the values of two waveforms at each instant to make a third waveform, of different shape; and the concept can be extended to the addition of the instantaneous values of many waves. The principle is illustrated on the following page.

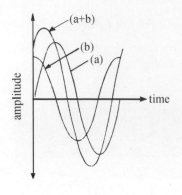

It is possible to show that any square or triangular waveform consists of the addition to a fundamental sine wave of an infinite number of its own harmonics. The distortions produced by non-linearity in amplifiers, in particular, can be analysed in this way.

If, however, a square or triangular wave is passed through a low-pass filter, the process is reversed and the higher order harmonics are removed from the input wave. When a 1 kHz square or triangular wave, for instance, is passed through a filter which attenuates all frequencies above 1.5 kHz, the resulting output in both cases will be a sine wave of 1 kHz frequency.

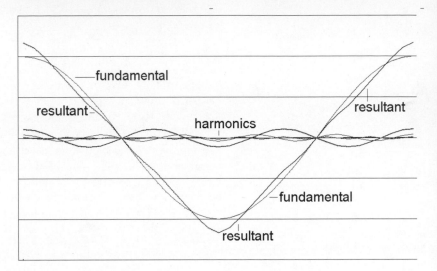

Figure 10.6   Addition of fundamental and odd harmonics to produce a roughly triangular waveform

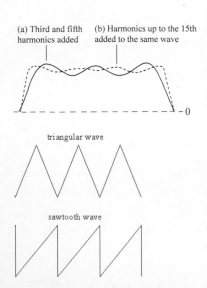

When a set of harmonics is added to a fundamental waveform, the amplitude of these harmonic waves will have an effect on the shape of the total waveform. Figure 10.6 shows half a cycle of a fundamental with the effect of adding odd harmonics (3rd, 5th, 7th) to a fundamental, amplitude A, with the amplitude of each harmonic inversely proportional to the square of its harmonic number. For example, we use A/9 of the 3rd harmonic, A/25 of the 5th harmonic and so on. The result, if this is taken to an infinite number of harmonics, is a triangular wave.

The illustration, left, shows the waveshapes which result from adding (a) the 3rd and 5th harmonics to a sine wave, and (b) the odd harmonics up to the 15th. The higher amplitudes of the harmonics are in this case inversely proportional to the harmonic number, so we use A/3 of the 3rd harmonic, A/5 of the 5th harmonic, etc., so that very high harmonics have little effect on the waveshape.

Note that rectangular waves contain high proportions of odd-numbered harmonics (3,5,7,9 ...), while triangular and sawtooth waveforms contain high proportions of the even-numbered harmonics. Pulse waveforms contain a mixture of both even and odd harmonics.

Wave analysis

The principle of *wave analysis* is important as a method of measuring amplifier distortion. A perfect sine wave signal applied to the input of an amplifier will produce a perfect sine wave output only if the amplifier causes no distortion. Any alteration in the waveshape brought about by distortion will cause harmonics to be present in the output signal. These harmonics can be isolated and measured to enable the amount of the distortion which has taken place to be calculated.

Filters

Filters are circuits that act on waveforms to change their amplitude, their phase, often both. Passive filter circuits which consist wholly of passive components will never increase the amplitude of a wave but will either reduce it or leave it unchanged. The gain of a passive filter, meaning the quantity:

$$\frac{\text{amplitude of output wave}}{\text{amplitude of input wave}}$$

is always unity or less. The amplitudes must, of course, be measured in the same way, so that if the input amplitude is measured as peak-to-peak, then the output amplitude must also be measured in this way.

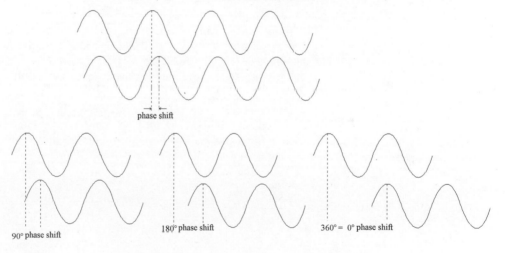

Figure 10.7 Phase shift

Filter circuits do not change the shape of sine waves, but they do alter the shapes of other waveforms (see later in this chapter). Many filter circuits cause the *phase* of a sine wave to shift. This means that the peak of the waveform at the output does not occur at the same time as the peak of the wave at the input (see Figure 10.7). Phase shift is measured in degrees, with 360° representing one complete cycle of difference between input and output waves. On this scale, 180° represents half a cycle of difference, and 90° a quarter of a cycle.

A double-beam oscilloscope, which uses two separate Y (vertical deflection) amplifiers and a common timebase to control two CRT beams, can be made to show phase shift by displaying both input and output waveforms on the same timebase. When the time difference t between the waves is measured, the phase shift is given in degrees by the expression: $t/T \times 360°$, where T is the time period or duration of a complete cycle of the waveform.

Example: The time difference between the peaks of two sets of waves is 10 μs, and the wave period of each is 50 μs. What phase shift has occurred, in degrees?

Solution: Substitute the data in the equation $t/T \times 360°$, then $\dfrac{10}{50} \times 360 = 72$, a phase shift of 72°.

Question 10.2

The phase shift between two waves is 45°. If the frequency of each wave is 5 kHz, what will be the time difference between peaks as measured by the oscilloscope?

Wave filters can be designed to cope with waves of any range of frequencies from low (a few Hz only) to high radio frequencies of many hundreds or even thousands of MHz. The components and circuit design of the filters vary considerably according to the frequency range they are designed to handle.

Whatever the range of frequencies handled by a filter, however, the main filter types are low-pass, high-pass, band-pass and band-stop. The symbols used in block and circuit diagrams to indicate these types of filter action are shown in Figure 10.8.

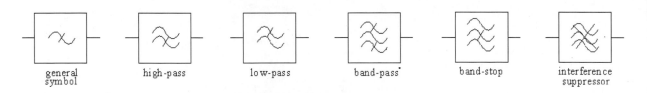

Figure 10.8   Filter symbols

A low-pass filter, as its name suggests, passes without attenuation the low frequencies of signals arriving at its input, but greatly reduces the amplitude of high frequency signals which are thereby heavily attenuated. Another way of putting it is to say that a low-pass filter has a pass-band of low frequencies and a stop-band of high frequencies.

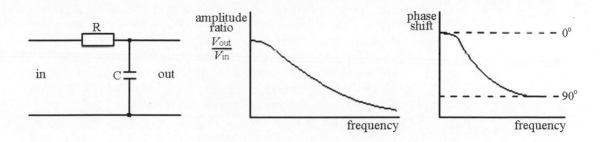

Figure 10.9 A simple lowpass filter and its performance graphs

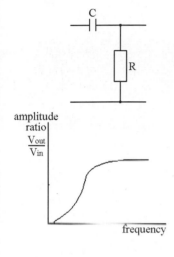

A simple form of low-pass filter is the RC filter shown in Figure 10.9 together with typical performance graphs. These graphs show the variation of both gain and phase plotted against frequency which the filter is capable of achieving.

The drawing, left, shows a single CR high-pass filter with its measured gain/frequency performance graph.

Simple RC filters do not give sufficient attenuation for most purposes, and filters which include inductors as well as capacitors and resistors are also required. Both the theory and the practical design of such filters present difficulties, but Figure 10.10 illustrates examples of both low-pass and high-pass filters.

When inductors and capacitors are combined in a filter circuit, band-pass and band-stop filters can be constructed. Band-pass filters will pass a predetermined range of frequencies without attenuation, but will attenuate other frequencies both above and below this range. A band-stop filter has the opposite action, greatly reducing the amplitude of signals in a given band but having little effect on signals outside it.

A further type of filter, perhaps less frequently encountered, is the all-pass filter. This type causes no attenuation of signals, but does bring about phase shifts from zero to 360°, over a wide range of frequencies.

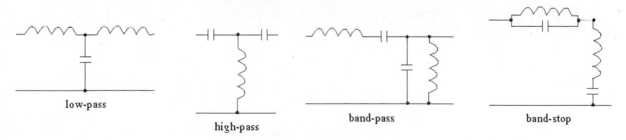

Figure 10.10 More complex filters

Wave shaping

The basic action of filter circuits on sine waves is, as has been seen, to attenuate and change the phase of the waves but to leave the shape of the wave unaltered. When the input wave to a circuit is not a sine wave, however, the action of even a passive circuit can change the shape of the wave (as well as its phase) considerably.

Circuits designed to change the shape of non-sinusoidal waves of this type are called wave shaping circuits. If active components (that is to say, components other than capacitors, inductors and resistors) are included, the shapes of sine waves themselves can also be changed.

Two important types of wave shaping circuits are the differentiating circuit and the integrating circuit. A differentiating circuit is one which has an output only when the input wave changes amplitude. The amplitude of its output depends on how fast the amplitude of the input wave itself changes: the faster the change, the greater the output.

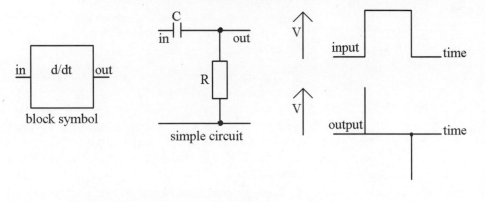

Figure 10.11   A differentiating circuit

The normal differentiating circuit will act as a high-pass filter for sine waves. Its effect on a square wave is shown in Figure 10.11. As the illustration shows, the effect is to emphasize the changes in the wave so that the output consists of a spike of voltage for each sudden change of voltage level of the square wave. The direction of the spike is the same as the direction of voltage change of the square wave.

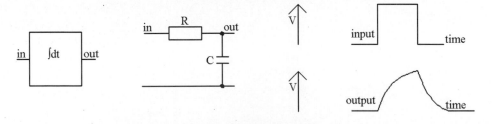

Figure 10.12   An integrating circuit

An integrating circuit removes sudden amplitude changes from a waveform, its effect being thus the opposite of that of the differentiating circuit. Integrating action on a square wave is shown in Figure 10.12. The circuit's action on a sine wave is that of a low-pass filter, smoothing out any sudden changes, the opposite of the action of the differentiating circuit.

When simple RC circuits are used for differentiating and integrating the circuit property called its *time constant* becomes important. As we have seen, it is measured by multiplying the value of R by the value of C. A differentiating circuit requires a time constant that is small compared with the wave period. An integrating circuit requires a time constant that is large compared with the wave period.

One result of these requirements is that differentiating or integrating circuits that are designed to handle a given wave frequency will have a different effect when a wave of either higher or lower frequency reaches them as an input.

The units in which the time constant is calculated must be strictly observed. If R and C are given in ohms and farads respectively, the time constant RC will appear in seconds. If (as is more likely) the units of R and C are given in kilohms and microfarads respectively, the constant RC will appear in milliseconds (ms). If R is in kilohms and C in nanofarads, RC will appear in microseconds (µs).

Example: Calculate the time constants of (a) 10K and 0.033 µF, (b) 470R and 0.1 µF, (c) 56K and 820 pF.

Solution:

(a) With R = 10K, C must be converted into nanofarads, 0.033 µF = 33 nF. The time constant is then $33 \times 10 = 330$ µs.

(b) R must be converted into kilohms. 470R = 0.47k. With C = 0.1 µF, the time constant is $0.47 \times 0.1 = 0.047$ ms, or 47 µs.

(c) With R = 56K, C must again be converted into nanofarads. 820 pF = 0.82 nF. The time constant is then $56 \times 0.82 = 45.9$ µs, or about 46 µs.

Except for a few low-frequency circuits, the unit of microseconds is the most convenient for use in time constant calculations.

Activity 10.1

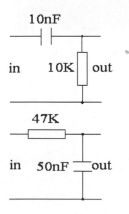

Construct the wave shaping circuits shown, left. For each circuit in turn, connect the oscilloscope to the output of the circuit and a square wave signal generator to the input. Calculate the time constant for the circuit. Now adjust the square wave generator until it produces a wave of 1 V p–p at 1 kHz. Sketch the output waveform from the wave shaping circuit.

Set the generator to a frequency of 100 Hz and repeat the measurement. Repeat the measurements using 100 Hz and 1 kHz sine waves and confirm that the only effect is alteration of phase shift.

Investigate also the effect on a sine wave input of a diode clipper circuit, as illustrated in Chapter 8.

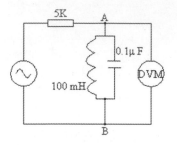

Activity 10.2

Construct the filter circuit illustrated in the drawing using a signal generator as the signal source and a digital voltmeter (DVM) as the detector. The inductor and capacitor should have values as indicated, using a polyester type of capacitor.

Start with the signal generator set to deliver 1 V output at a frequency of 1 kHz. Read the output of the voltmeter and if the reading is too small for the lowest range of the instrument, increase the output of the signal generator until a reading can be taken. Now increase the frequency in steps of 100 Hz, taking a reading of output at each frequency up to 2000 Hz. Plot your readings on a graph of amplitude (Y-axis) against frequency.

When you have plotted the graph, find by adjusting the signal generator the frequency for which the output amplitude is a maximum, and confirm that this could also be deduced from your graph.

Now rearrange the circuit so that the inductor and the capacitor are in series between points A and B. Take readings of amplitude and frequency once again between 1 kHz and 2 kHz, and plot a graph of the results. How does the action of the series LC circuit compare with the action of the parallel circuit?

Activity 10.3

Connect the inductor and the capacitor used in Activity 10.2 in series between points A and B, and repeat the actions of Activity 10.2.

Answers to questions

10.1   0.4 m or 40 cm.

10.2   25 µs.

11 Input and output transducers

Switches and contacts

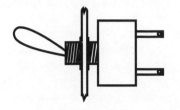

Switches have a low-resistance between contacts in the ON setting, and a very high-resistance in the OFF setting. The value of resistance when the switch is on (*made*) is called the *contact resistance*. The amount of the contact resistance depends on the area of contact, the contact material, the amount of force that presses the contacts together, and also in the way that this force has been applied. A wiping action provides a lower resistance than a simple pressing action.

If the contacts are scraped against each other in a wiping action while they are forced together, then the contact resistance can often be much lower than can be achieved when the same force is used simply to push the contacts straight together. In general, large contact areas are used only for high-current operation and the contact areas for low-current switches, as used for electronics circuits, will be small. The actual area of electrical connection is always less than the physical area of the contacts, because contacts are never precisely flat. The usual solution to this problem is to use a multiple-contact structure.

A switch contact can be made entirely from one material, or electroplating can be used to deposit a more suitable contact material. By using electroplating, the bulk of the contact can be made from any material that is mechanically suitable, and the plated coating will provide the material whose resistivity and chemical action is more suitable. In addition, plating makes it possible to use materials such as gold and platinum which would make the switch impossibly expensive if used as the bulk material for the contacts. Contacts for switches are therefore normally constructed from steel or from nickel alloys, with a coating of material that will supply the necessary electrical and chemical properties for the contact area. The usual choice of materials is the same as is used for relays with the addition of copper and beryllium-copper.

Ratings, a.c. and d.c.

Switch ratings are always quoted separately for a.c. and for d.c., with the a.c. rating often allowing higher current and voltage limits, particularly for inductive circuits. When d.c. flowing through an inductor is decreased, a reverse voltage is induced across the inductor, and the size of this voltage is equal to inductance multiplied by rate of change of current. The effect of breaking the inductive circuit is a pulse of voltage, and the peak of the pulse can be very large, so that arcing is almost certain when an inductive circuit is broken unless some form of arc suppression is used. The circuits for arc suppression will be noted later; they use either a capacitor to absorb the energy of the pulse, or a diode to divert the current.

Activity 11.1

Use a multimeter to test the continuity of switch and relay contacts. Check the effect of operating the switch/relay.

Contact arrangement

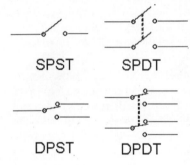

SPST   SPDT

DPST   DPDT

Switch contact configurations are primarily described in terms of the number of *poles* and number of *throws* or *ways*. A switch pole is a moving contact, and the throws or ways are the fixed contacts against which the moving pole can rest. The term *throw* is usually reserved for mains switches, mostly single- or double-throw, and *way* for signal-carrying switches. A single-pole, single-throw switch (SPST) will provide on/off action for a single line, and is also described as single-pole òn/off. Such switches are seldom used for a.c. mains nowadays and are more likely to be encountered on d.c. supply lines.

For a.c. use, safety requirements call for both live and neutral lines to be broken by a switch, so that double-pole single-throw (DPST) switches will be specified for this type of use. Double-throw switches are not so common for mains switches in electronics use – their domestic use is in two-way switching systems, and they have some limited applicability for this type of action in electronics circuits. A double-throw switch will also be described as a changeover type. The word *throw*, used to describe the number of fixed contacts, is a reminder of the snap-over action that is required of this type of switch. The single-pole, double-throw switch (SPDT) is also called a single-pole change-over (SPCO), and the double-pole, double-throw switch is also called double-pole change-over (DPCO). Contact arrangements are shown diagrammatically, left.

Where a single or multiple contact is required to make connection to more than two fixed contacts per pole, then the fixed contacts are referred to as poles. This type of configuration is much more common in signal switches, and switches of up to twelve ways per pole are available in the traditional wafer form.

The configuration of a double-throw switch also takes account of the relative timing of contact. The normal requirement is for one contact to break before the other contact is made, and this type of break-before-make action is standard. The alternative is make-before-break (MBB), in which the moving pole is momentarily in contact with both fixed contacts during the changeover period. Such an action is permissible only if the voltages at the fixed contacts are approximately equal, or the resistance levels are such that very little current can flow between the fixed contacts. Once again, this type of switching action is more likely to be applicable in signal-carrying circuits. There may be a choice of fast or slow contact make or break for some switches.

A switch may be *biased*, meaning that one position is stable, and the other, off or on, is attained only for as long as the operator maintains pressure on the switch actuator. These switches are used where a supply is required to be on only momentarily – often in association with the use of a hold-on relay – or can be interrupted momentarily. A biased switch can be of the off-on-off type, in which the stable condition is off, or the on-off-on type, in which the stable condition is on. The off-on-off type is by far the more common. The ordinary type of push-button switch is by its nature off-on-off biased, though this is usually described for such switches as momentary action (see, however, alternate action below), and biased switches are also available in toggle form.

Arcing

Arcing is one of the most serious of the effects that reduce the life of a switch. During the time of an arc very high temperatures can be reached both in the air and on the metal of the contacts, causing the metal of the contacts to vaporize, and be carried from one contact to the other. This effect is very much more serious when the contacts carry d.c., because the metal vapour will also be ionized, and the charged particles will always be carried in one direction.

Arcing is almost imperceptible if the circuits that are being switched run at low voltage and contain no inductors, because a comparatively high voltage is needed to start an arc. For this reason, then, arcing is not a significant problem for switches that control low voltage, such as the 5 V or 9 V d.c. that is used as a supply for solid-state circuitry, with no appreciable inductance in the circuit. Even low voltage circuits, however, will present arcing problems if they contain inductive components, and these include relays and electric motors as well as chokes. Circuits in which voltages above about 50 V are switched, and particularly if inductive components are present, are the most susceptible to arcing problems.

Temperature range

The normal temperature range for switches is typically –20°C to +80°C, with some rated at –50°C to +100°C. This range is greater than is allowed for most other electronic components, and reflects the fact that switches usually have to withstand considerably harsher environmental conditions than other components. The effect of very low temperatures is due to the effect on the materials of the switch. If the mechanical action of a switch requires any form of lubricant, then that lubricant is likely to freeze at very low temperatures. Since lubrication is not usually an essential part of switch maintenance, the effect of low temperature is more likely to be to alter the physical form of materials such as low-friction plastics and even contact metals.

Flameproofing

Flameproof switches must be specified wherever flammable gas can exist in the environment, such as in mines, in chemical stores, and in processing plants that make use of flammable solvents. Such switches are sealed in such a way that sparking at the contacts can have no effect on the atmosphere outside the switch. This makes the preferred type of mechanism the push-on, push-off type, since the push button can have a small movement and can be completely encased along with the rest of the switch.

Connections

Switch connections can be made by soldering, welding, crimping or by various connectors or other screw-in or plug-in fittings. The use of soldering is now comparatively rare, because unless the switch is mounted on a PCB which can be dip-soldered, this will require manual assembly at this point. Welded connections are used where robot welders are employed for other connection work, or where military assembly standards insist on the greater reliability of welding. By far the most common connection method for panel switches, as distinct from PCB mounted switches, is crimping, because this is very much better adapted for production use. Where printed circuit boards are prepared with leads for fitting into various housings, the leads will often be fitted with bullet or blade crimped-on connectors so that switch connections can be made.

Transducers
Terms

A transducer is a device that converts energy from one form into another. In this book we are concerned only with transducers that have an electrical input or output. *Sensitivity* or *efficiency* refers to the fraction of input energy which is converted to output energy, and for many transducers this can be very low, of the order of a few per cent. The percentage efficiency is often masked by quoting the sensitivity in terms of the amount of output per unit input.

The electrical *impedance* of transducers can range from a fraction of an ohm to many megohms, and it affects the type of connections that can be made to the transducer. In addition, if the transducer has an electrical output, the impedance level usually determines the typical electrical output level.

Linearity means the extent to which the output is proportional to the input, so that a perfectly linear device would have a graph of output plotted against input that would be a straight line.

Note that a device termed a *sensor* is a form of transducer intended for measurement and control actions. A sensor need not be efficient in terms of the amount of energy converted, but it must have a linear or otherwise predictable response.

> The responsivity is defined as:
> **output signal/input signal**
> The detectivity is defined as:
> **S/N of output signal/size of input signal**
> where S/N has its usual electrical meaning of signal to noise ratio. This latter definition can be reworked as:
> **responsivity/output noise**
> if this makes it easier to measure.

Two important measurable quantities can be quoted in connection with any sensor or transducer. These are *responsivity* and *detectivity*, and though the names are not necessarily used by the manufacturer of any given device, the figures are normally quoted in one form or another. Responsivity is a measure of transducing efficiency if the two signals are in comparable units (both in watts, for example), but it is normally expressed with very different units for the two signals.

Input transducers

microphone
symbol

The most familiar transducer for sound energy into electrical energy is the microphone, and microphone types are classified by the type of transducer they use. The overall sensitivity is expressed as millivolts or, more usually, microvolts of electrical output per unit intensity of sound wave. In addition, though, the impedance of the microphone is also important.

A microphone with high impedance usually has a fairly high electrical output, but the high impedance makes it very susceptible to hum pick-up. A low impedance is usually associated with very low output, but hum pickup is almost negligible.

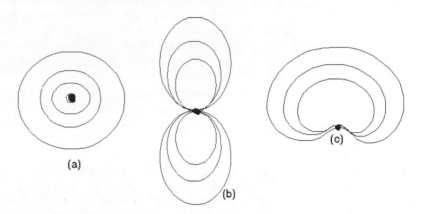

Figure 11.1 Simplified response curves for microphone types. (a) Omnidirectional, (b) velocity-operated, (c) cardioid (heart-shaped)

A microphone must be connected to a matching impedance. If a high impedance microphone is connected to a low-impedance amplifier input there will be losses in signal resulting in feeble output.

Another important factor is whether the microphone is *directional* or *omnidirectional*. If the microphone operates by sensing the *pressure* of the air in the sound wave, then the microphone will be omnidirectional, picking up sound arriving from any direction. If the microphone detects the *velocity* (speed and direction) of the air in the sound wave, then it is a directional microphone, and the sensitivity has to be measured in terms of direction as well as amplitude of the sound wave. The microphone types are known as pressure or velocity operated, omnidirectional or in some form of directional response (such as cardioid).

The frequency response of a microphone is usually very complicated because the housing of the transducer unit will act as a filter for the sound waves, causing the response to show peaks and troughs unless the housing is very carefully designed.

Microphone types

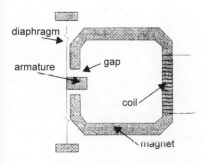

Microphone types in common use include moving iron (dynamic), moving-coil, crystal, and electret capacitor types.

The principle of the moving iron (dynamic, or variable reluctance) microphone is illustrated in the cross-section view. A powerful magnet contains a soft-iron armature in its magnetic circuit, and this armature is attached to a diaphragm. The magnetic flux in the magnetic circuit alters as the armature moves. A coil wound around the magnetic circuit at any point will give a voltage which is proportional to each change of magnetic flux, so that the electrical wave from the microphone is proportional to the acceleration of the diaphragm.

The linearity of the conversion can be reasonable for small amplitudes of movement of the armature but very poor for large amplitudes. Both the linearity, and the amplitude over which linearity remains acceptable can be improved by appropriate shaping of the armature and careful attention to its path of vibration. These features depend on the maintenance of close tolerances in the course of manufacturing the microphones, so that there will inevitably be differences in linearity between samples of microphones of this type from the same production line. The output level from a moving-iron microphone can be high, of the order of 50 mV, and the output impedance is fairly high, typically several hundred ohms. Magnetic shielding is always needed to reduce mains hum pick-up in the magnetic circuit.

The moving-coil microphone uses a constant-flux magnetic circuit in which the electrical output is generated by moving a small coil of wire in the magnetic circuit. The coil is attached to a diaphragm, and as before, the maximum output occurs as the coil reaches maximum velocity between the peaks of the sound wave so that the electrical output is at 90 degrees phase angle to the sound wave. The coil is usually small, and its range of movement very small, so that linearity is excellent, impedance is low and the signal output is also low.

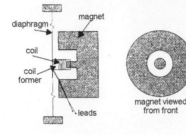

Activity 11.2

Test the continuity of a moving-coil microphone using a multimeter. Test also the continuity of the coil of a moving-coil loudspeaker. Note also the direction of cone movement relative to the polarity of the test leads.

Crystal (piezoelectric) microphones use a piezoelectric crystal element either alone or (not so commonly nowadays) connected to a diaphragm. The crystal, a material such as barium titanate, is one in which the arrangement of atoms ensures that any tiny displacement of the atoms due to vibration will cause a voltage to be generated across the crystal, and this is sensed by depositing conducting films across opposite faces of the crystal, to which the output leads are connected. The impedance level is of the order of several megohms, as distinct to a few ohms for a moving-coil type, and the output is in the high millivolt range rather than microvolts.

The capacitor microphone, left, is now used almost exclusively in electret form. An electret is the electrostatic equivalent of a magnet, a piece of insulating material which is permanently charged. A slab of electret is therefore the perfect basis for a capacitor microphone in which the vibration of a metal diaphragm near a fixed charge causes voltage changes between the diaphragm and a metal backplate. This allows very simple construction for a microphone, consisting only of a slab of electret metallized on the back, a metal (or metallized plastic) diaphragm, and a spacer ring, with the connections taken to the conducting surface of the diaphragm and of the electret. This is now the type of microphone which is built into cassette recorders, and even in its simplest and cheapest versions is of considerably better audio quality than the piezoelectric types that it

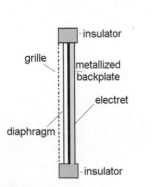

displaced. The use of a FET preamp solves the problem of the high impedance of this type of microphone.

Magnetic tape

tapehead
symbol

Magnetic tape uses the principle of hard magnetic material. Early tape recorders in 1898 used steel tape, but it was not until 1941 that the BASF company hit on the idea of coating a thin plastic tape with magnetic powder, using iron oxide. The recording head is made from a soft magnetic material with low hysteresis so that its magnetic flux will follow the fluctuations of a signal current through the winding. The same head is also used for replaying tape, using the varying flux from the moving tape to induce currents in the head. The symbol for a tapehead is illustrated, left.

Because iron oxide is a hard magnetic material, it can be magnetized by a strong magnetic field close to the tape, and this magnetism will be retained when the tape is removed from the field. When the magnetized tape is drawn past a coil, variations of magnetism in the tape will induce varying voltages in the coil, so providing the replay action. One head is normally used for both recording and replaying, with switched connections either to the output of an amplifier or to the input of the amplifier. A separate head is fed with the output from a high-frequency (typically 100 kHz) oscillator to act as an erase head. The erase head is operated only during recording.

See Chapter 16 for a more complete description of tape recorder principles.

Activity 11.3

Test the continuity of the recording head of a tape/cassette recorder, using a multimeter. Note that the head may need to be demagnetized after this measurement – use a head demagnetizer if available.

Output transducers
Earphones and loudspeakers

Earphones have been in use for considerably longer than microphones or loudspeakers, since they were originally used for electric telegraphs. The power that is required is in the low milliwatt level, and even a few milliwatts can produce considerable pressure amplitude at the eardrum – often more than is safe for the hearing.

The methods that are used for earphones and for loudspeakers are essentially the same as are used for microphones, because the action is (unusually) reversible. Symbols are illustrated, left.

The first type of successful earphone transducer was a moving-iron type, and this principle has been extensively used ever since for both earphones and (earlier in the century) for loudspeakers. As applied to the telephone, the earphone uses a magnetized metal diaphragm so that the variation of magnetization of the fixed coil will ensure the correct movement of the diaphragm. The unit is sensitive but the linearity is poor and moving iron units are seldom used.

The use of moving-coil earphones has been less common in the past, but these are now in widespread use thanks to the popularity of miniature cassette players. As applied to earphones, moving-coil construction permits good linearity and controllable resonances, since the amount of vibration is very small and the moving-coil unit is light and can use a diaphragm of almost any suitable material.

A variation of the moving-coil principle that has been successfully used for earphones is the electrodynamic (or orthodynamic) principle. This uses a diaphragm which has a coil built in, using printed-circuit board techniques. The coil can be a simple spiral design, or a more complicated shape (for better linearity), and the advantage of the method is that the driving force is more evenly distributed over the surface of the diaphragm.

The piezoelectric principle has also been used for earphones in the form of piezoelectric (more correctly, pyroelectric) plastics sheets which can be formed into very flexible diaphragms.

A *loudspeaker* has to be housed in a cabinet whose resonances, dimensions and shape will considerably modify the performance of the loudspeaker unit. The assembly of loudspeaker and cabinet will be placed in a room whose dimensions and furnishing are outside the control of the loudspeaker designer, so that a whole new set of resonances and the presence of damping material must be considered. The transducer of a loudspeaker system is sometimes termed the 'pressure unit', and its task is to transform an electrical wave, which can be of a very complex shape, into an air-pressure wave of the same waveform. To do this, the unit requires a *motor unit*, transforming electrical waves into vibration, and a diaphragm which will move sufficient air to make the effect audible.

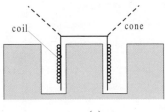

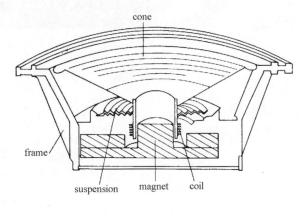

(a)

(b)

Figure 11.2   Moving-coil loudspeaker. (a) Cross-section of coil and magnet, (b) cutaway view of complete unit

Most loudspeakers use the *moving-coil* principle, Figure 11.2. The coil is wound on a former at the apex of a stiff diaphragm which is held in flexible supports so that the coil is in an intense magnetic field. A signal current through the coil will then vibrate the diaphragm, producing sound waves. To be really effective, a loudspeaker has to be housed in a cabinet (enclosure) whose design assists the loudspeaker to move a large volume of air, acting like a transformer for air waves. Even in a fairly good enclosure, the efficiency of a typical hi-fi loudspeaker is very low, around 1 per cent.

Another problem is that no single loudspeaker (a full-range unit) can achieve good linearity over the whole audio range of frequencies. The usual method of dealing with this is to use two or more units in a single enclosure, with a high-frequency unit (tweeter) and a bass unit (woofer) dealing with the extremes of the audio range. A filter called a crossover unit separates the electrical signals so that the appropriate frequency range is fed to the appropriate loudspeaker unit.

The moving-coil loudspeaker has a nominal value of impedance, typically 4 ohms, but a graph of impedance plotted against frequency is very complicated in shape, with peaks and dips caused by the mechanical resonances of the loudspeaker components and the enclosure.

Miniature loudspeakers can generally handle only the higher frequencies, but hi-fi headphones are miniature loudspeakers in enclosures that are designed to achieve a reasonable level response over a large part of the audio frequency range.

Ultrasonic wave transducers are used for sending or receiving ultrasonic signals through solids or liquids, and can operate in either direction if required. The important ultrasonic transducers are all piezoelectric or magnetostrictive, because these types of transducers make use of vibration in the bulk of the material, as distinct from vibrating a motor unit which then has to be coupled to another material. These transducers are used to drive ultra-sonic cleaning tanks.

Magnetostriction is the change of dimensions of a magnetic material as it is magnetized and demagnetized. Several types of nickel alloys are strongly magnetostrictive, and have been used in transducers for the lower ultrasonic frequencies, in the range of 30–100 kHz.

A magnetostrictive transducer consists of a magnetostrictive metal core on which is wound a coil. The electrical waveform is applied to the coil, whose inductance is usually fairly high, so restricting the use of the system to the lower ultrasonic frequencies. The core magnetostriction will cause vibration, and this will be considerably intensified if the size of the core is such that mechanical resonance is achieved.

The main use of magnetostrictive transducers has been in ultrasonic cleaning baths, as used by watchmakers and in the electronics industry. The *piezoelectric* transducers have a much larger range of application, though the power output cannot approach that of a magnetostrictive unit.

The piezoelectric transducer crystals are barium titanate or quartz, and these are cut so as to produce the maximum vibration output or sensitivity in a given direction. The crystals are metallized on opposite faces to provide the electrical contacts, and can then be used either as transmitters or as receivers of ultrasonic waves. The impedance levels are high, and the signal levels will be millivolts when used as a receiver, a few volts when used as a transmitter.

Question 11.1

Explain how ultrasonic cleaning operates and state the typical range of frequencies employed.

A *surface acoustic wave* (SAW) device is an arrangement that uses two transducers back to back with electrical input and electrical output but an acoustic wave (which is usually ultrasonic) between. A signal input to one transducer causes the wave to be set up, and the wave reaching the second transducer causes an electrical output at the same frequency, but delayed compared to the original signal.

The main application is to filtering. If the transducers are formed on a crystal which has a mechanical resonance, the output will be strongly frequency dependent. By careful mechanical design, band-pass filters can be constructed which have a very steep transition between pass-band and stop band, very much better than can ever be achieved with any LC combination, active or passive.

D.c. permanent magnet motor

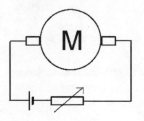

The d.c. permanent-magnet motor, whose symbol and typical connections are illustrated, left, is used extensively in equipment that needs mechanical actions, such as video recorders, CD players, etc. As the name implies, the motor uses a rotating armature with a winding connected through a commutator, with a field supplied by a permanent magnet.

This type of motor is favoured because its speeds can be controlled by controlling the current through the armature winding, and the direction of rotation can be reversed by reversing the direction of current through the armature winding. The armature current can be supplied from a circuit incorporating a power transistor, so that the current can be closely regulated to provide precise speed control.

Relays

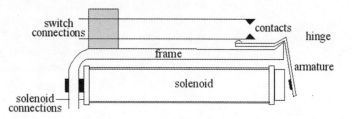

Figure 11.3   The classic type 8000 relay construction

The classical form of relay (type 8000) is illustrated in Figure 11.3. The coil, usually a solenoid, is wound on a former around a core, and a moving armature forms part of the magnetic circuit. When the coil is energized, the armature moves against the opposition of a spring (or the elasticity of a set of leaves carrying the electrical contacts) so as to complete the magnetic circuit. This movement is transmitted to the switch contacts through a non-conducting bar or rod, so that the contacts close, open or change over depending on the design of the relay. The contacts are subject to the same limits as those of mechanical switches, but the magnetic operation imposes its own problems of make and break time, contact force and power dissipation. Like so many other electronics components, relays for electronics use have been manufactured in decreasing sizes, and it is even possible to buy relays that are packaged inside a standard TO-5 transistor can.

Question 11.2

How may switch contacts be protected when switching inductive loads?

The relay has to be specified by the operating voltage and current, and also by the switching arrangements. They are commonly classed as normally open (NO), meaning that applying power to the relay will close the contact(s), or change-over (CO) meaning that applying power will change over the switch connections. Normally-closed connections are less usual, and applying power to such a relay will open the contacts. Contacts are usually single or double pole, so that typical arrangements are SPNO, SPCO, DPNO or SPCO (also known as DPDT). Circuit diagrams often use the detached contact method. The coil of the relay is shown in one part of the diagram (in the collector circuit of a transistor, perhaps), and the switching contacts are in another part of the diagram. This avoids the need to show long and confusing connecting lines on a diagram.

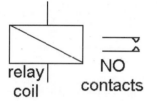

relay coil   NO contacts

Activity 11.4

Measure the resistance of a relay operating coil, using a multimeter.

Cathode-ray tubes

The cathode-ray tube or CRT is an active component which nowadays is just about the only device you are likely to come across that uses the same principles as the old-style radio valves. The CRT is used to convert variations of voltage into visible patterns, and is applied in instruments (oscilloscopes), for television, and for radar.

The three basic principles are that:

- Electrons can be released into a vacuum from a red-hot cathode (a metal coated with metal-oxides).

- These electrons can be accelerated and their direction of movement controlled by using either a voltage between metal plates or a magnetic field from a coil that is carrying an electric current.

- A beam of electrons striking some materials such as zinc sulphide will cause the material (called a phosphor) to glow, giving a spot of light as wide as the beam.

Oscilloscope CRT

The electrons from the cathode are attracted to the positive anode by an accelerating voltage of several kV. On their way, they have to pass through a pinhole in a metal plate, the *control grid*. The movement of the electrons through this hole can be controlled by altering the voltage of the grid, and a typical voltage would be some 50 V *negative* compared to the cathode. At some value of negative grid voltage, the repelling effect of a negative voltage on electrons will be greater than the attraction of the large positive voltage at the far end of the tube, and no electrons will pass the grid – this is the condition we call *cut-off*.

Electrons that pass through the hole of the grid can be formed into a beam by using metal cylinders at a suitable voltage. By adjusting the voltage on one of these cylinders, the *focus electrode*, the beam can be made to come to a small point at the far end of the tube. This end is the screen, and it is coated with a material (a *phosphor*) that will glow when it is struck by electrons. The phosphor is usually coated with a thin film of aluminium so that it can be connected to the final accelerating (*anode*) voltage. The whole tube is pumped to as good a vacuum as is possible; less than a millionth of the normal atmospheric pressure.

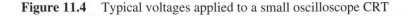

This arrangement will produce a point of light on the centre of the screen, and any useful CRT must use some method of moving the beam of electrons. For small CRTs a set of four metal plates can be manufactured as part of the tube and these deflection plates will cause the beam to move if voltages are applied to them. The usual system is to arrange the plates at right angles, and use the plates in pairs with one plate at a higher voltage and the other at a lower voltage compared to the voltage at the face of the tube. This system is called *electrostatic deflection*.

Monochrome electrostatic CRT

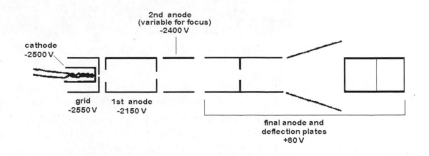

Figure 11.4   Typical voltages applied to a small oscilloscope CRT

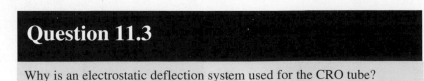

Question 11.3

Why is an electrostatic deflection system used for the CRO tube?

A typical electrostatically deflected CRT is intended for oscilloscope use, and will use a screen size of, typically 5 to 7 inches diameter. Rather than earthing the cathode and connecting the anode to several kV voltage, the tube is used with the anode earthed, and the cathode at a high negative voltage. This makes it possible to connect the deflection plates directly to a modest voltage that can be supplied from transistor circuits. The other voltages for focus and accelerating electrodes can be supplied from a resistor network connected between the earth and the most negative voltage (applied to the grid). Figure 11.4 shows typical voltages.

Magnetic deflection

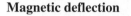

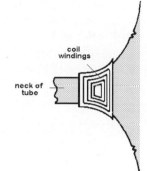

coil windings

neck of tube

There is an alternative method for deflecting the electron beam which is used for larger tubes, particularly for computer monitors, radar and TV uses. A beam of electrons is a current flowing through a vacuum, and magnets will act on this current, deflecting the beam. The easiest way of doing this is to place coils around the neck of the tube and pass current through these coils to control the beam position on the face of the tube. This magnetic deflection method is better suited for large CRTs such as are used for simple radar, CCTV or computer monitor applications. The coil windings are usually shaped so that they can be placed over the curved portion of the neck of the tube, as illustrated here.

The form of deflection that is most common for CRTs is a *linear sweep*. This means that the beam is taken across the screen at a steady rate from one edge, and is returned very rapidly (an action called *flyback*) when it reaches the other edge. A sawtooth waveform is needed to generate such a linear sweep. An electrostatic tube can use a sawtooth voltage waveform applied to its deflection plates, and a magnetic-deflection can use a sawtooth current applied to its deflection coils. The difference is important, because the electrostatic deflection requires only a sawtooth voltage with negligible current flowing, but the magnetically deflected tube requires a sawtooth *current*, and the voltage across the deflection coils will not be a sawtooth, because the coils act as a differentiating circuit. In fact, the voltage waveform is a pulse, and this is used in TV receivers to generate a very high voltage for the CRT. Unit 6 deals with colour cathode-ray tubes and their use in TV receivers.

Answers to questions

11.1   By vibrating the dirt particles so that they fall away from the component being cleaned. Frequency range is 35–45 kHz.

11.2   By shunting the contacts with capacitor or resistor–capacitor circuits to absorb the back-emf pulse.

11.3   Because it produces a more linear deflection and this aids the precision of measurement. It is also small and lighter than the alternative magnetic deflection system, making the CRO more portable.

12   Electronic modules

Amplifiers, oscillators and filters

$$G_v = \frac{\text{output signal voltage}}{\text{input signal voltage}}$$

$$G_i = \frac{\text{output signal current}}{\text{input signal current}}$$

$$G_p = \frac{\text{output signal power}}{\text{input signal power}}$$

A transformer can have voltage gain or current, but not both, so that it has no power gain.

An amplifier is a device that has input and output terminals. A signal at the input terminals will appear with greater amplitude at the output terminals. If this larger-amplitude signal is in all other respects, especially frequency and waveshape, an exact copy of the signal at its input terminals, the amplifier is a *linear* amplifier. Any type of circuit that operates with signals of different amplitudes is an analogue circuit – compare this type of circuit with the digital type.

When the output signal has greater voltage amplitude than the input signal, the amplifier is said to have *voltage gain* (G_v). This gain is defined as the ratio shown here with both amplitudes measured in the same way.

When the output signal has greater current amplitude than the input signal, the amplifier is said to have *current gain* (G_i). This gain is defined as the ratio shown here, again with both measured in the same way.

When an amplifier produces both voltage and current gain, its output signal will have greater power than its input signal. The *power gain* (G_p) of the amplifier is defined as shown here, and this is also equal to voltage gain x current gain. so that $G_p = G_v \times G_i$.

These gain figures are often converted into decibel form, see Chapter 2. An amplifier is represented on a block diagram by the type of arrowhead symbols shown in Figure 12.1. The output terminal is always assumed to be at the tip of the arrowhead.

An inverting amplifier has an output that is the inverse of the input, this is often described as a 180° phase shift. The non-inverting amplifier has an output that is in phase with the input.

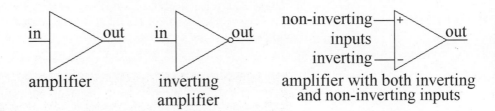

Figure 12.1   The symbols for an amplifier. The symbol at the right-hand side is also used for an op-amp

Decibel scales

Some ratios need to be expressed in a way that is more in keeping with our senses. For example, doubling the size (amplitude) of a sound wave does not make it seem twice as loud. For comparing some measurements, then, we use the decibel (dB) scale, as noted in Chapter 2. You should revise the definition of decibel scales at this point.

A decibel scale is normally used when the gain of an amplifier is plotted against frequency, see later. The decibel corresponds to the smallest change that we can detect when we are listening to the amplifier.

Classes of amplifier circuit

Audio frequency (AF) amplifiers handle the range of frequencies lying within the approximate range 30 Hz to 20 kHz. This is about the range of frequencies which can be detected by the (younger) human ear. Such amplifiers are used in record players and tape recording, in the sound section of TV receivers, in cinema sound circuits and in many industrial applications. Frequencies up to about 100 kHz, well beyond the range of the human ear, are often for convenience included in the AF range of a signal generator.

AF voltage amplifiers have fairly high voltage gain, and medium to high values of output resistance. AF power amplifiers give large values of current gain. They have very low output resistance so can pass large signal currents into low resistance loads such as loudspeakers.

Fault conditions are indicated by low gain (or no gain at all), by waveform distortion, by reduced bandwidth, or by changes in the input or output resistances.

D.c. amplifiers, as their name suggests, are used to amplify d.c. voltages or very low frequency signals. A d.c. amplifier with a voltage gain of 100 (40 dB), for example, could produce an output of 1 V d.c. from an input of 10 millivolts d.c.

D.c. amplifiers are used in industrial electronic circuits such as photocell counters, in measuring instruments such as strain gauges, and in medical electronics (where they could be used to measure, for example, the electrical voltages in human muscles).

The main problem with d.c. amplifiers is their stability. It is difficult to ensure that their d.c. output does not vary, or drift, from one minute to the next, or change as room temperature changes. Voltage drift is thus the main fault of d.c. amplifiers, but variations in gain and a type of instability which causes either low frequency or high frequency oscillations are also sources of operating trouble.

Wideband amplifiers are amplifiers that provide the same value of gain over a large range of frequencies typically from d.c. to several MHz. One important special application is as signal amplifiers in oscilloscopes; but it is true to say that most equipment used to make measurements on signal voltages will include a wideband amplifier.

No form of tuning is possible for such amplifiers; and their design is complicated by the requirement that they must provide the same values of voltage gain for d.c. as they do for high frequency a.c. signals. Common faults include instability of bandwidth, variations in gain, and the danger of oscillation.

Video amplifiers are a type of wideband amplifier used for the specific purpose of amplifying signals to be applied to cathode ray tubes (CRTs). The video signal is the signal which carries all picture information for TV or radar sets, and the video amplifier is the circuit used for amplifying it. A frequency range from d.c. to 5.5 MHz is desirable in video amplifiers used in TV, but a wider range is required to handle most radar signals.

Detailed amplifier circuits are covered in the Level-3 syllabus. In this chapter, only the measurable features of amplifiers are discussed, not details of their circuitry.

Distortion

pure sine wave

clipped wave

crossover distortion

If the waveshape at the output of the amplifier is not a true copy of the waveshape at the input, the amplifier is said to be producing *distortion*. Distortion can be caused by overloading i.e. by putting in a signal input whose peak-to-peak voltage is greater than the amplifier can accept. If signal input is normal, distortion generally indicates a fault in the amplifier circuit itself.

Two types of distortion are particularly noticable. Clipping means that the tips of a sine wave are not amplified, and this can affect one tip of a wave or both. Crossover distortion is caused by an amplifier that does not amplify the lowest voltage levels in a wave, causing a sine wave to form a small step just as it passes through the zero voltage level.

Clipping is caused by overloading an amplifier, meaning that the input wave amplitude is too large for the amplifier to handle. Crossover distortion is a fault of incorrect transistor bias.

Not all amplifiers are intended to be linear, however. Pulse amplifiers, for example, are not intended to preserve in every respect the shape of the wave they amplify. A pulse is used for timing, and the leading edge of the pulse is the usual reference that is used. The one thing a pulse amplifier must not do is alter the slope of the leading edges of the pulse inputs. Providing that the leading edge is sharp, the amplitude of the pulse is less important, and the shape of the portion of the pulse following the leading edge is also less important.

Measurements

Voltage, current and power gains, all defined above, are three important measurements that can be made on any amplifier. To measure the gain of an amplifier, a small signal is injected into the input and the amplitude of the output signal is measured (see Figure 12.2, following). For most purposes, voltage gain is the figure of most interest, and it is the voltages of the input and output signals which have to be measured. Suitable input signals are obtained from a signal generator fitted with a calibrated attenuator; a circuit which reduces the signal by a given amount (e.g. ÷10, ÷100, ÷1000). The output signal is generally measured against the Y (vertical deflection) amplifier calibration of a cathode-ray oscilloscope (CRO). Voltage gain is usually expressed in decibels (dB).

A common problem is that the input signal from the signal generator may be too small to be measured by the oscilloscope even when the oscilloscope is set at its maximum sensitivity. In practical work, for example, a signal as low as 30 mV p–p might have to be measured; but few of the oscilloscopes used in servicing can reliably measure such a quantity. To overcome the difficulty, the attenuator of the signal generator is switched so that the signal is made ten times greater (in this example, to 300 mV, or even one hundred times greater at 3 V).

Question 12.1

An amplifier has an input of 5 mV and an output of 2.5 V, both measured peak to peak. What is its gain in decibels?

Quantities such as these can be readily measured with the CRO; and when this has been done, the attenuator is reset to reduce the signal generator output to the value corresponding to a 30 mV input. Some signal generators provide a full-voltage output for the oscilloscope and an attenuated output for the signal to the equipment under test.

When both input and output signals have been measured, the voltage gain is found by use of the formula:

$$G = \frac{\text{output voltage amplitude}}{\text{input voltage amplitude}};$$

the most convenient units for measurement with the CRO are peak-to-peak. Whatever method is used must be used for both input and output measurements.

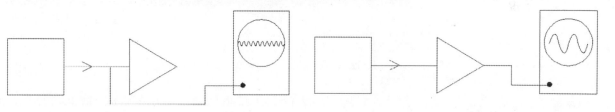

(a) oscilloscope measuring amplitude of input signal   (b) oscilloscope measuring amplitude of output signal

Figure 12.2   Measuring voltage gain

Similar methods are used to measure the *bandwidth* of an amplifier. This is the range of frequencies over which the amplifier can be effectively used. The upper and lower limits of bandwidth are taken to be the frequencies at which the voltage gain of the amplifier is reduced to 70.7 per cent of its value (−3dB) in the middle of the frequency range. The frequencies where the gain is 70.7 per cent of maximum are often called the 3 dB points, see Chapter 2.

As well as the change of gain, the *phase* of the output signal will change as the frequency is altered. At mid-frequency, the phase of the output will be either 0° or 180°, meaning that it is at the same phase as the input or inverted. Note that inverted and 180° phase have the same meaning only for sine-waves. By the time that the gain has fallen to 70.7 per cent, the phase will have changed also, by an angle of 45°.

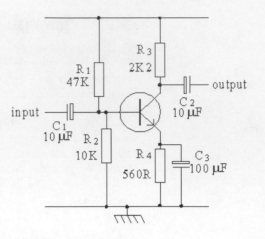

Figure 12.3   Diagram for Activity 12.1

Activity 12.1

Construct the voltage amplifier shown in Figure 12.3, using a BC107 or BFY50 transistor. Connect the amplifier to a 9 V supply and check that the collector voltage falls within the range 4 V to 5 V. Use the oscilloscope to set the sine wave of the signal generator to 30 mV p–p at 400 Hz.

Connect the output of the signal generator to the input of the amplifier, and the output of the amplifier to the oscilloscope. Measure the peak-to-peak amplitude of the signal from the output of the amplifier. Now calculate the voltage gain from the formula shown above.

The bandwidth of an amplifier is expressed in one of two ways: by quoting either the two limiting frequencies or the difference between them. Thus the bandwidth of an audio amplifier may be quoted as, for example, '40 Hz to 25 kHz', or the bandwidth of a radio frequency amplifier as '10 kHz centred on 465 kHz'. Specifically, the bandwidth is defined as the frequency range between the two 3 dB or *half-power* points, which corresponds to the points where the voltage gain falls to 70.7 per cent of its maximum value (and the power to half of its maximum value).

Bandwidth is assessed by plotting the actual voltage gain of the amplifier at a range of different frequencies. A graph of gain plotted against frequency (Figure 12.4) then shows the two frequencies (the 3 dB points) at which the gain has fallen to 70.7 per cent of mid-band gain (3 dB down).

The fact that the power gain is half of the mid-band power gain at the 3 dB point is a convenient way of defining bandwidth for power amplifiers, so that you will see the phrase *half-power bandwidth* used in connection with power amplifiers.

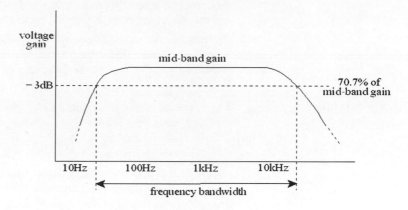

Figure 12.4   A typical graph of bandwidth

Note that the X axis of the graph along which frequency is measured is calibrated on a logarithmic scale in order to accommodate the wide range of frequencies covered by a typical amplifier bandwidth.

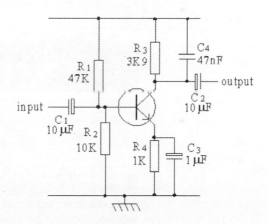

Figure 12.5   Diagram for Activity 12.2

Activity 12.2

In the circuit shown in Figure 12.5, measure voltage gain at 50 Hz, 100 Hz, 200 Hz, 1 kHz, 2 kHz, 4 kHz, 8 kHz, 12 kHz and 20 kHz. On logarithmic graph paper (linear by 4-cycle log) plot a graph of voltage gain against frequency for the amplifier in question (which you will find has purposely been given component values which will give it a narrow bandwidth). From this deduce the bandwidth.

Now, using the same frequency range, supply a square wave input of the same amplitude. Sketch the output waveshape for each different input signal frequency. Comment on the results.

Multistage amplifiers

When several stages of amplification and/or attenuation are connected in series, the total gain is $G_1 \times G_2 \times G_3$, where G_1 is the gain of the first stage, G_2 is the gain of the second stage, and G_3 is the gain of the third stage. If one of these stages is an attenuator, the gain figure will be a fraction.

If the gain is quoted in dB, then the total gain in dB will be $G_{1dB} + G_{2dB} + G_{3dB}$, and, if one stage is an attenuator its dB figure will be negative. For example, if the first stage has voltage gain of 100 (40 dB) and the last stage has a gain of 80 (38 dB), with a potential divider that passes ½ of the signal (−6 dB), then the dB gains are: $40 - 6 + 38 = 72$ dB.

Distortion measurement

Amplifier distortion is measured using distortion meters. A pure sinewave input is fed to the amplifier, and the output is applied to the distortion meter which will detect any harmonics of the wave, indicating distortion. The amount of distortion is often quoted in terms of total harmonic distortion (THD), and for a hi-fi amplifier you could expect this figure to be less than 0.1 per cent.

Question 12.2

An amplifier stage with an output resistance of 2K2 is coupled to a stage with input resistance 1K. Each stage has a voltage gain of 50. What is the overall gain in dB? HINT – draw a block diagram first.

Oscillators

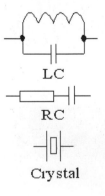

LC

RC

Crystal

An oscillator is an amplifier that provides its own input by using positive feedback. That is, a signal is taken from the output and fed back through some network of components into the input *in phase* so as to maintain the output signal. The feedback network usually consists of frequency-determining components as shown. There are two conditions that have to be met to sustain oscillations:

- the amplifier gain must be greater than the attenuation of the feedback circuit and,
- the feedback must be positive.

If the loop gain is only just greater than 1, then the output waveshape will be sinusoidal. If the gain is much greater than 1, oscillations will be violent and this results in pulse or square waveshapes. The frequency of the oscillations depends upon the values of the components in the feedback loop and the output amplitude depends largely on circuit component and supply voltage values. The essential function of an oscillator is to convert the d.c. power supplied into it into a.c. power as the output signal and like all energy converting devices, it dissipates heat due to its inefficiencies.

RC and **crystal-controlled oscillators** can be used for a wide range of frequencies, but the RC type is used mainly for audio frequencies, though its range can be extended to several MHz. The most common type of RC oscillator is the Wien bridge circuit, which overcomes the problems of obtaining a pure sine wave from RC oscillators.

Pulse (or *astable* or *aperiodic*) **oscillators** also contain a network which acts as a time control, plus an amplifier. The amplifier is, as is usual in oscillator circuits, connected so that its input is provided from its own output signal by positive feedback; but the operating principle on which it works is different from that of a sinusoidal oscillator.

In all pulse oscillators, the amplifier is in operation for only a very short time in any cycle, often for much less than a microsecond. For the remainder of the cycle, the waveform is generated by the changing level of the voltage across a capacitor as a resistor feeds current into the capacitor or drains current out of it. It is only the steeply sloping sides of the waveform which are caused by amplifier action. All the remaining parts of the waveform are generated by the resistor capacitor network. Depending on the design of the oscillator, the output can be square waves, pulses or triangular (sawtooth) waves. Remember that a pulse can be obtained by differentiating a square wave and a triangular wave can be obtained by integrating a square wave.

Question 12.3

An amplifier is faulty and oscillates at a frequency that can be shown on an oscilloscope. What steps might you take to stop the oscillation?

Activity 12.3

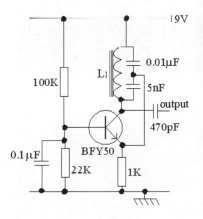

Construct the oscillator circuit shown here. The circuit is a sinewave oscillator of a type called a Colpitts oscillator. It generates a sinewave at a frequency of around 1 MHz. The coil L_1 consists of 25 turns of wire occupying a width of 11 mm on a former of 7 mm diameter, with a ferrite core.

When the circuit has been built and checked, connect the output to the oscilloscope and connect a 9 V supply to the oscillator supply terminals. Measure the amplitude and frequency of the output.

Activity 12.4

The circuit for this activity is a square-wave oscillator of a type called an astable or multivibrator. Do not dismantle the multivibrator circuit when you have finished experimenting with it, but retain it for use in a later exercise. When the circuit has been built and checked, connect either output to the oscilloscope and connect a 9 V supply to the oscillator supply terminals. Measure the amplitude and frequency of the output. The action of this circuit is described in Chapter 14.

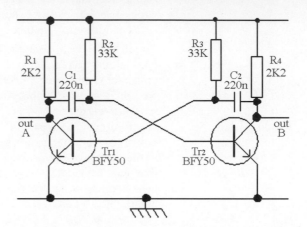

Figure 12.6   Diagram for Activity 12.4

* Note that an oscillator, operating correctly, should provide an output signal whenever a d.c. supply is connected to it. A faulty oscillator may be revealed by failure to produce an output signal (or at best an intermittent one) or by generating an output waveform of distorted waveshape, incorrect frequency or insufficient amplitude.

Answers to questions

12.1   53.97 dB, 54 dB when rounded up.

12.2   57.8 dB.

12.3   Reduce the gain of the amplifier for the frequency of oscillation or find the path of the positive feedback loop and change the phase of the feedback signals.

Unit 5

13 Combinational logic

Number systems

1, 2, 3, 4, 5, 6, 7, 8, 9, A, B, C, D, E, F, 10, 11, 12, 13, 14, 15, 16, 17, 18, 19, 1A, 1B, 1C, 1D, 1E, 1F, 20.........

0, 1, 10, 11, 100, 101, 110, 111, 1000, 1001, 1010, 1011, 1100, 1101, 1110, 1111, 10000, 10001, 10010, 10011, 10100, 10101, 10110, 10111, 11000, 11001, 11010, 11011, 11100, 11101, 11110, 11111............

Power	of TEN	of TWO
0	1	1
1	10	2
2	100	4
3	1 000	8
4	10 000	16
5	100 000	32
6	1 000 000	64
7	10 000 000	128
8	100 000 000	256
9	1 000 000 000	512

Number conversion

Ordinary counting is done using ones and tens, it is a *denary* system. In all number systems, the number that is raised to these different powers, like the denary 10, is called the *base*, and the numbers that are used for the powers are called the *exponents*.

Ten, however, is not the only number which can be used as the base for a system of counting. Bases other than 10 can be used without any change in the method of writing the exponents. In the scale of 8, for example, the number 163 means 3 ones, plus 6 eights, plus one $8^2 (= 64)$ making a total of $64 + 48 + 3 = 115$ in denary terms. Such a scale is used in some computing work, as is a scale of 16. In this latter *hexadecimal* scale, as it is called, we need to be able to express the denary numbers 10, 11, 12, 13, 14 and 15, and the first six letters of the alphabet (A, B, C, D, E, and F respectively) are used for the purpose.

Another counting system is the binary scale. As its name binary implies the base is 2. The distance to the left of the suppressed point is used to show what power of two is being used, and the only digits that are used are 0 and 1. Thus the binary number 1101 is the equivalent of a denary number calculated as follows: the first place to the left of the suppressed binary point represents 2^0, or 'ones'. There is a 1 in that place, so it must be counted. The second place to the left of the point represents 2^1, or 'twos'. There is a zero in this place, so nothing is added to the count. The third place represents 2^2, or 4; and a 1 in this position means that 4 must be added to the count just as a 1 in the fourth place to the left ($2^3 = 8$) means that 8 must be added. The full denary equivalent is therefore $1 + 0 + 4 + 8 = 13$.

Binary figures to the right of the binary point are used to express binary fractions, such as $\frac{1}{2}, \frac{1}{4}, \frac{1}{8}, \frac{1}{16}$ (2^{-1}, 2^{-2}, 2^{-3}, 2^{-4}) and so on to further inverse powers of 2.

The table, left, shows powers of up to 9 for both denary and binary systems. This illustrates that binary numbers must contain more digits than their denary equivalent. One digit in a hexadecimal number is the equivalent of four binary digits, so that the hexadecimal scale is more compact then the binary scale, and conversions to and from hexadecimal are simpler.

Denary numbers are converted to binary numbers by the following process. Divide the denary number by 2, and write down any remainder at the head of a separate remainder column. Divide what is left by 2, and again write down any remainder in the remainder column, immediately under the first entry. Clearly, all entries in this remainder column must be either 1s (if there is a remainder after a division by 2) or 0s (if there is not).

Continue this process of division by two until your original denary number is reduced to 1. Now since 'two into one doesn't go'; you put in a zero as the last entry in the diminishing number column, and the remainder 1 as the last entry in the remainders column. Then read off the remainder column starting from the bottom and moving upwards, and write down the ensuing string of 0s and 1s, from left to right. The left-most bit is described as the 'most significant bit' (MSB) and the right-most as the 'least significant bit' (LSB).

Example: Convert the denary number 527 to binary.

Solution:

Action	Diminishing Number	Remainder
Divide 527 by 2	263	1
$\div 2$	131	1
$\div 2$	65	1
$\div 2$	32	1
$\div 2$	16	0
$\div 2$	8	0
$\div 2$	4	0
$\div 2$	2	0
$\div 2$	1	0
$\div 2$	0	1

Start from the bottom of the remainder column and write out the digits in it from left to right. The binary number corresponding to denary 527 then appears as: 1000001111.

The conversion of binary numbers to denary is much simpler if this method is used. Write out the binary figure, for example 100101. Ignore all the zeros in that figure, and concentrate on the correct power of 2 for each of the 1s according to its position in the binary number. Then add the denary figures on the bottom line and the full denary equivalent of binary 100101 appears as $32 + 4 + 1 = 37$.

$$
\begin{array}{cccccc}
2^5 & & 2^2 & & 2^0 \\
1 & 0 & 0 & 1 & 0 & 1 \\
32 & & 4 & & 1
\end{array}
$$

Binary arithmetic

Adding binary numbers is carried out in the same way as for denary numbers except that the rule is now that two 1s add to 'zero and 1 carried forward', while three 1s added together equal '1 and 1 carried forward'. For example:

$$
\begin{array}{r}
1011 \\
+1101 \\
\hline
=11000
\end{array}
$$

In the units column the two 1s add up to binary 10 (spoken as 'one zero', not 'ten'). A zero is written down and 1 is carried forward. The next column then appears as $1 + 1 + 0$, which therefore again means that a zero is written down and 1 is carried forward. The same thing happens in the third column, but in the fourth the carry forward makes the total $1 + 1 + 1$. So a 1 is now written down, leaving the carry forward 1 to appear alone in the fifth column on the left.

Binary subtraction can be done by the familiar 'borrow' method that is used in denary subtraction; but an easier method is available which is called '2s complement'. Write down the large number, and below it the smaller number 'complemented'. This means that every 0 in the smaller number becomes a 1, and every 1 a 0. Next, add to this complemented binary number the unit 1, and the result is added to the larger number. The figure on the extreme left of this sum is disregarded if it would make the number of digits larger, and what is left is the result of the binary subtraction.

Note that both the binary numbers in this method of subtraction must have the same number of digits. If they do not, you must add the appropriate number of zeros to the left of the smaller number.

Example: Subtract binary 110 from binary 1101.

Solution: Write down the larger figure .... .1101

The smaller number must be rewritten as 0110

Complement this binary number... 1001

Add 1 to give... .1010

Now add 1101 to make 10111

Discard the left-hand digit, and the binary remainder is................... .0111

This remainder is the equivalent of denary 7, which is indeed the remainder when binary 110 (= denary 6) is subtracted from binary 1101 (= denary 13).

A quick method for 2's complement is to copy down the digits from the right, up to and including the first 1, then invert each other bit in turn. Try this in the example.

The fact that the '2's complement' method involves only addition makes it the method used for computing, because it avoids the need to use a different set of circuits for subtraction.

Question 13.1

What binary number do you get when you subtract 10011 from 11100?

The peculiar advantage of the binary scale for electronic counting systems is that only the figures 0 and 1 need to be represented. It is much easier to design a circuit in which a transistor is biased either fully on or fully off than it is to design one in which varying values of collector voltage represent different figures in the denary scale. A binary system needs no carefully calculated bias voltages, and will not be greatly affected by changes in either supply voltage or component values. In addition, the recording and replaying of binary numbers is much less affected by faults in the recording process.

In a normal system of digital logic, '1' is represented by a comparatively high voltage say 3 V, 5 V, 10 V, 12 V or whatever the supply voltage happens to be and '0' by a much lower voltage typically around 0 to 0.2 V. This system is called positive logic. It is also possible to represent 1 by a low (or negative) voltage and 0 by a higher voltage, a system called negative logic. We usually refer to these voltages only in terms of the digits 0 and 1.

- The binary system described here is also known as 8-4-2-1 binary to distinguish it from other forms of binary code such as Grey code.

Another way of using binary code is in the form of *binary-coded decimal* (BCD). In a BCD number, four binary digits (or bits) are used to code one denary number, using the binary equivalents of denary 0 to 9. For example, the denary number 815 can be coded in BCD as 1000 0001 0101, the binary codes for the three denary digits. Arithmetic in BCD is very complicated, but the code is very useful when values have to be displayed.

Logic gates

A *gate* is a circuit which allows a signal to pass through when it is open but blocks the signal when it is closed. The opening and closing of the gate is done by means of other electrical signals. A *logic gate* uses only digital signals of the 0 or 1 type, and the aim of all logic gate circuits is to ensure that the output becomes 1 only when some fixed combination of inputs present themselves to the gate. A gate is also called a *combinational circuit*. Figure 13.1 shows two logic actions carried out by switch circuits.

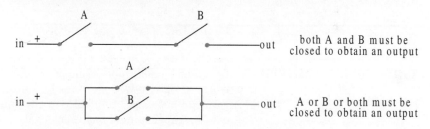

Figure 13.1 Switch circuits for logic actions

The simplest gate uses only one input and is the NOT gate or inverter. If the input is 0 then the output is 1. If the input is 1 then the output is 0. The simplest types of two input gates are called AND, OR, NAND and NOR gates according to their action. The action of a gate is shown in its *truth table* which is a list of the outputs that are obtained from every possible combination of inputs into the gate. A truth table is much easier to understand than the circuit diagram of a gate, which, when an IC is used may not even be available.

The truth tables for all AND, OR, NAND and NOR gates are shown here, left. These show that the output of the AND gate is 1 *only* when both inputs are 1 (i.e., only when both Input A and Input B are at logic 1). The output of the OR gate is 1 when *either* Input A *or* Input B is at logic 1, or when both inputs are at logic 1.

The NAND and NOR gates are no more than the inverses of the AND and the OR gates respectively. Check carefully through their inputs and outputs in the tables to see what this implies.

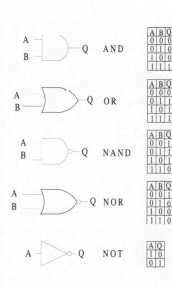

A	B	Q
0	0	0
0	1	0
1	0	0
1	1	1

AND

A	B	Q
0	0	0
0	1	1
1	0	1
1	1	1

OR

A	B	Q
0	0	1
0	1	1
1	0	1
1	1	0

NAND

A	B	Q
0	0	1
0	1	0
1	0	0
1	1	0

NOR

A	Q
1	0
0	1

NOT

- Conventionally, we illustrate gate action with reference to two inputs, though gates with larger numbers of inputs are manufactured and used.

The standard actions of the four types of gate mentioned are represented on diagrams by the symbols shown in Figure 13.2. Note that the British Standard (BS) symbols (seldom used) differ from the international symbols which will be found in most logic diagrams. Both sets of symbols should be known. The international symbols are sometimes described as the 'United States Military Specification' or Milspec.

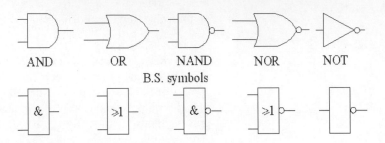

Figure 13.2   Logic gate symbols

Gate combinations

By suitable connection of the appropriate standard gates, any type of truth table can be achieved. Conversely, since the truth tables of all the standard gates are known, a truth table for any combination of gates can be worked out. The procedure for analysing a gate circuit is as follows:

1   Prepare a series of columns for the logic gate combination whose truth table you wish to work out, and head them A, B, C, etc., for as many original inputs as the combination is to receive.

2   Prepare another set of columns for all the intermediate points in the system where the output of one gate forms the input (or inputs) to another gate (or gates). Label these columns with further consecutive letters of the alphabet, such as D, E, etc.

3   Label a final column Q, for the final output.

4   Fill in the first set of input columns (A, B and C in the truth table, see following) with a series of binary numbers starting with 000, 001, 010, 011 and so on, until the row is reached in which the numbers are all 1s.

5   Then work out and fill in for every D, E, etc., column the output which the intermediate gate in question would give if it received the inputs displayed in columns A, B and C.

6   Complete every row by working out and filling in the final output Q.

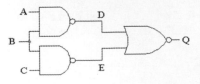

The diagram, for example, shows a circuit consisting of two NAND gates followed by a NOR gate. It has three inputs A, B, C, two intermediate stages D, E and an output Q. Six columns are thus needed.

The A, B, C columns are first filled in, with all the binary numbers from 000 to 111 so providing for every possible combination of inputs. The

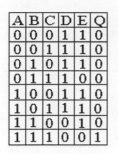

A	B	C	D	E	Q
0	0	0	1	1	0
0	0	1	1	1	0
0	1	0	1	1	0
0	1	1	1	0	0
1	0	0	1	1	0
1	0	1	1	1	0
1	1	0	0	1	0
1	1	1	0	0	1

intermediate values in columns D and E are then entered from the truth table for the NAND gate (which you will remember shows a 0 output only if both inputs are at 1. Otherwise the output is 1). Finally, the Q column is filled in from the truth table for the NOR gate, as illustrated.

- This particular logic gate gives an output only for the combination of inputs shown. These inputs might represent either (say) the number 7 (binary 111), or a series of signals from safety switches which would permit a machine to be switched on only when certain combinations of switches were pressed.

Activity 13.1

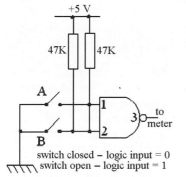

switch closed – logic input = 0
switch open – logic input = 1

See the diagram, left. The SN7400 integrated circuit contains four NAND gates, each with two inputs, and is available as a 14-pin DIL pack. This can be conveniently mounted on a solderless breadboard for experimental work.

With the IC suitably mounted, connect Pin 14 to the positive terminal of a regulated 5 V supply, and Pin 7 to the negative terminal of the supply. Then connect a voltmeter (using its 10 V range) with its positive lead to Pin 3, and its negative to earth. Connect switches between earth and Pins 1 and 2 as shown, labelling the open circuit position '1' and the s/c position '0'. The reason for labelling the switches thus is that closing a switch will result in a very low voltage, so connecting the input to logic zero; while opening the switch will connect the pin through the 47K resistor to + 5 V, which is logic 1. Label the switches A and B, and fill in the truth table for the NAND gate.

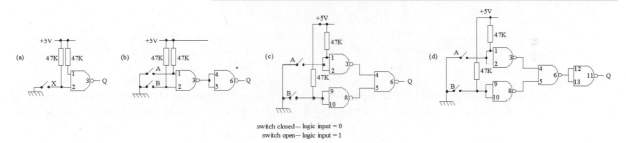

switch closed— logic input = 0
switch open— logic input = 1

Figure 13.3 Diagram for Activity 13.2

- Each input of a gate should be taken to either +5 V or 0 V, and never allowed to float. This rule can be relaxed for d.c. operation, but not when pulses are used.

Activity 13.2

Having completed Activity 13.1, work out truth tables for the other connections of gates shown, with pin numbers, in Figure 13.3. These correspond to other standard gate types which are named at the end of this chapter, but do not consult them until you have really tried to compile your own. The results of this exercise should be noted, because they show how NAND gates can be used to provide the action of other gates. Work out for yourself what other gate types are represented by the truth tables.

NOR gates can also be used in combination to provide the output of any other gate, and this equivalence was worked out in theory in De Morgan's theorem some 150 years ago. The practical implication is that digital circuits can be constructed entirely from one simple gate, either NAND or NOR, making the construction of digital ICs easier.

Digital IC families

Digital ICs were the first IC types to be manufactured, and several methods for forming gates and other digital circuits have been used over the years. Ignoring early types such as DTL and RTL, the most enduring form of IC has been TTL, with the letters meaning transistor–transistor logic. This type of IC uses no resistors in its internal design, only transistors, and the inputs are taken to the emitters of a transistor that has more than one emitter. The original type of TTL system (now called STTL) operates with a regulated voltage supply of +5 V (which must never exceed 5.25 V). Any voltage between 0 and +0.8 V will be taken as logic 0 level, and any voltage between +2 V and +5.5 V will be taken as logic 1 level. Voltage inputs that lie between +0.8 V and +2.0 V must be avoided, because they could be taken as either 0 or 1.

STTL is now manufactured only for replacement purposes, and later versions such as Schottky (LSTTL) are now current. The LSTTL types operate using the same inputs, but have much lower power dissipation. In addition, there are several MOS types of ICs, some of which can use a wide range of operating voltage, others being used where low power dissipation is important. The table below summarizes important characteristics for some of these IC families.

The most important digital gate circuits nowadays are those that combine a very short delay time (fast-acting gates) with very low power, and in this respect the MOS types are much superior to the older bipolar types.

The HC and AC CMOS types have for most purposes replaced the earlier TTL types, and carry the same numbers, so that, for example, 74C00 (CMOS) carries out the same logic as 7400 (TTL). The older CMOS types used the 4000 series of numbering.

Logic circuits must use a regulated power supply that cannot apply excessive voltage to any unit. Logic circuits also need to be protected from excessive or reverse voltage inputs, and this can be done using diodes. For MOS IC circuits, these protection diodes are built into the chips.

Table 13.1 IC families summarized

	STTL	LSTTL	CMOS	74HC	74AC
V+ supply	5 V	5 V	3–15 V	5 V	5 V
Imax/1	40 µA	20 µA	10 pA	10 pA	10 pA
Imax/0	–1.6 mA	–0.4 mA	10 pA	10 pA	10 pA
Imax/out	16 mA	8 mA	1 mA*	4 mA	4 mA
Delay	11–22 ns	9–15 ns	40–250 ns*	10 ns*	5 ns
Power	10 mW	2 mW	0.6 µW	1 µW	1 µW
Frequency	35 MHz	40 MHz	5 MHz	40 MHz	100 MHz

NOTES:

$V+$ supply = normal positive supply voltage level

Imax/1 = maximum input current for logic level 1

Imax/0 = maximum input current for logic level 0

Imax/out = maximum output current

Delay = propagation delay in nanoseconds, a low delay means a fast device

Power = No-signal power dissipation per gate in mW or µW

Frequency = Typical operating frequency

1 pA $= 10^{-12}$ A

* These quantities depend on the supply voltage level.

Practical measurements and faultfinding

The instruments that are used for measurement and testing in digital circuits include the familiar types that are used also for work on analogue circuits. Measurement of steady voltage levels, such as power supply voltage, can be carried out using multimeters, either analogue (pointer) or digital (number display) types.

The d.c. supply voltage level is very important for digital circuits, because a low supply voltage will result in incorrect voltages for level 1, causing 1s to be counted as 0s in the digital ICs. This is therefore the first point to check when a digital circuit ceases to work or starts to work erratically.

You should know how the d.c. supply is regulated, because failure of a regulator chip may be the cause of a low or absent supply voltage. If, however, a regulator has failed in such a way as to cause the supply voltage to rise, it is very likely that this may have destroyed all the ICs, depending on the type used. In this respect, the CMOS ICs are more rugged than some TTL types.

Some circuits incorporate both over-voltage and reverse voltage protection, so that power is shut off in the event of regulation failing, or a battery being connected the wrong way round.

Logic circuit action can be carried out using d.c. meters if the logic inputs can be obtained from switches rather than from fast-acting circuits. This makes fault-finding much simpler, because you can observe the inputs to each gate and check that the outputs are as expected.

The CRO is also used in digital servicing where live signals are used, because a suitable type can check that a clock pulse is available. Note, however, that modern computer servicing needs oscilloscopes of a very high standard because clock frequencies are very high, typically 200 MHz or more.

The most common types of instruments that are used specifically for working digital circuits (and not for analogue circuits) are logic probes and monitors. A logic probe, as the name suggests, is held on to one line of a logic circuit, and its indicator will show if the line voltage is high, low or pulsing. This can be used to check if the expected signal is present on a line, and if the expected signal is not found you can then check for continuity along the line and for correct operation of the device that controls the line.

Logic monitors are an extension of a probe, consisting of a set of probes that can be held on a set of lines or clipped to the pins of a chip.

For advanced servicing work, automatic diagnostic equipment, such as that available from *Polar Instruments*, is unbeatable for difficult problems. Such instruments are by no means easy to use, and require a fair amount of operator training.

Answer to question

13.1   01001, which is denary 9.

Answers to questions, Activity 13.2

(a) NOT   (b) AND   (c) OR   (d) NOR

14   Multivibrators, timers and logic circuits

Multivibrators and timers

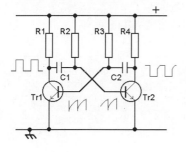

Either transistors or logic gates can be used to make the circuits we call *multivibrators*. These are classed as *astable*, *monostable* or *bistable* according to the way that they work, but the basis is always of two units that are cross-coupled, with the output of one fed to the input of the other.

The circuit illustrated here is an astable, meaning that it oscillates continually. The frequency of oscillation depends on the values of the time constants, formed here by the capacitors and resistors marked as C1, C2, R2 and R3. The output of this circuit is approximately a square wave, and by using more advanced circuitry, the shape of the wave can be almost a perfect square shape, with very low rise and fall times.

A circuit like this can be used to provide timing signals for other digital actions, and it can be crystal controlled so that the timing is very precise. Such a circuit is called a *clock*, since it sets the timing for the other circuit attached to it. Clocked circuits are particularly important for computing.

When a cross-coupled circuit contains one direct coupling and one CR coupling, it is a *monostable*. The diagram, following, illustrates a monostable circuit, again formed by using transistors since this is simpler to follow than the more usual IC types.

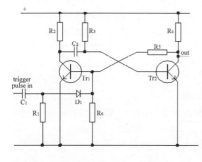

The monostable has one stable state, in this example with the second transistor Tr_2 conducting heavily because its base is connected to the positive supply through resistor R_1. When a brief positive pulse turns on transistor Tr_1, this will also turn off Tr_2, allowing the collector voltage to rise, and this state will persist until the capacitor C_2 charges and allows Tr_2 to conduct again. The circuit will then rapidly return to its original stable condition. This circuit is used to produce a pulse of a set width (set by the values of C_2 and R_3) from any brief positive input pulse.

Question 14.1

What effect would the choice of transistor type have on an astable multivibrator circuit?

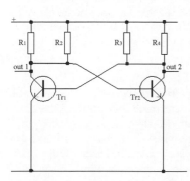

The *bistable* is illustrated, again in transistor form, in the diagram, left. The circuit can be stable with either Tr_1 or Tr_2 conducting, but is totally unstable if both are conducting. In this simple basic circuit, there is no provision for using a single pulse input, but in a more elaborate circuit (not illustrated here) a pulse at the input will be directed by *steering diodes* to one of the transistors, switching it over and so also switching the other. The next pulse will reverse the process, because of the altered bias on the steering diodes, returning the bistable to its original state. Two pulses in will therefore produce one pulse out. Bistables are seldom constructed from separate transistors because the IC form is more convenient.

Using logic gates as oscillators

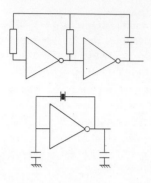

Logic gates can be used to construct oscillator circuits, with the advantages that they can operate at very high speeds, and provide signals that conform to the standard TTL levels. The drawings, left, show a simple oscillator using RC timing, and one that uses crystal control. The principle is exactly the same as that of analogue circuit oscillators, using a timing circuit as part of the positive feedback loop that causes the oscillation.

Logic ICs are used to construct oscillators used for clock pulses (clock oscillators). A clock pulse is used to synchronize the actions of logic circuits, an example will be illustrated in more detail later.

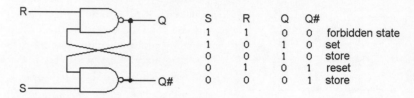

S	R	Q	Q#	
1	1	0	0	forbidden state
1	0	1	0	set
0	0	1	0	store
0	1	0	1	reset
0	0	0	1	store

Figure 14.1   A bistable made from NAND gates

NOTE: Read Q# as NOT Q

Figure 14.1 illustrates a bistable constructed from NAND gates. The sequence table for this circuit is also illustrated. Because the Q# output is, by convention, always required to be the opposite of the Q output, the input conditions S = 1, R = 1 must never be used. Note that in this circuit, the inputs R = 0, S = 0 causes the circuit to store an output at the value obtained when one input was 0 and the other 1. This storing action is also called latching. Latches can be obtained in IC form, and it is never necessary to construct such circuits from gate ICs.

The 555 timer

Though it is not strictly a digital circuit, the type 555 timer is used extensively as a clock generator (astable) where precise frequency is not important, or as a timer (monostable) circuit.

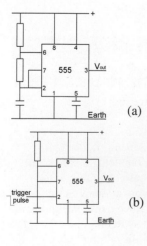

An external CR series circuit is used to set the time constant, and an input to the timer will generate a change of voltage at the output. This voltage will change back to its original values after a time determined by the time constant. The 555 can be used as an astable (a) or a monostable (b), and these basic circuits are illustrated, left. A more recent version, the 8555, will run from power supplies as low as 1.5 V and operates at higher frequencies.

The chip pin layout for the usual 8-pin DIL form is shown in Figure 14.2, using pins 1 and 8 for earth and positive supply respectively. The output is from pin 3, and the other pins are used to determine the action of the chip. Of these, pins 2, 6 and 7 are particularly important. A negative-going pulse on pin 2 will start the timer action. Pin 7 provides a discharge current for a capacitor that is used for timing, and pin 6 is a switch input that will switch over the output of the circuit as its load voltage changes. For most uses of the chip, these pins are connected to a CR circuit whose charge and discharge determines the time delay or the wave-time of the output. In the diagram of Figure 14.2 the dotted outline encloses the internal circuits of the 555 and the components and dotted connections outside illustrate the connections of a capacitor and resistor used to determine timing.

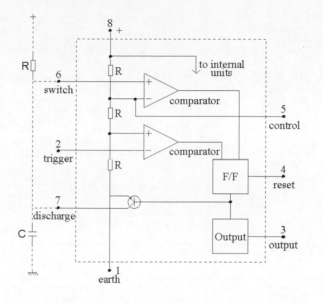

Figure 14.2   The pin connections and diagram of the 555 timer

Logic systems
Frequency division and counting

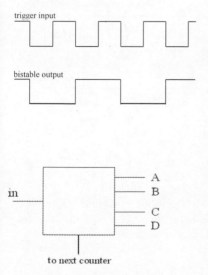

Frequency division is possible using the type of *bistable* circuit whose basic action is illustrated, left. As the diagram shows, the output only moves from 0 to 1 at every second 1-to-0 change of the voltage at the input. If the input to this circuit is a square wave, the output will be a square wave at half the frequency of the input square wave. This is a binary counter, a very important type of circuit.

The output of such a divider stage can of course be taken to the input of another similar stage so as to achieve another halving of frequency. In this way any given frequency can be divided down by any power of two.

If the input to a chain of divider circuits such as that described is a train of pulses with the number of pulses controlled by a gate, the divider chain will also act as a binary *counter*. After a number of pulses have been received at the input, the output of every bistable in the chain will be at either 0 or 1, representing a binary digit. Because every divider divides by 2, every output digit represents a power of 2, so that the outputs form a binary number equal to the number of input pulses received. Used in this way, the binary divider chain becomes a binary counter.

The most common form of single IC counter circuit is the BCD (binary coded decimal) counter, illustrated left with its sequence table. Input pulses up to denary 9 are counted, and binary outputs can be obtained for four outputs (2^3, 2^2, 2^1, 2^0). On the tenth pulse, all the outputs reset to zero, and

pulses	A	B	C	D
0	0	0	0	0
1	0	0	0	1
2	0	0	1	0
3	0	0	1	1
4	0	1	0	0
5	0	1	0	1
6	0	1	1	0
7	0	1	1	1
8	1	0	0	0
9	1	0	0	1

an output pulse becomes available to carry over to a second counter stage (2^7, 2^6, 2^5, 2^4) if another stage is needed.

The outputs from counters such as the BCD counter can be displayed as decimal figures on seven-segment readouts. A *decoder* circuit is needed (consisting of logic gates) to ensure that a given binary set of outputs produces the correct decimal number on the readout. The decoder is usually an integrated circuit; and IC chips forming combined counter–decoders, and even counter–decoder–displays, are nowadays available.

By way of illustration of digital techniques, Figure 14.3 shows the block diagram of a simple *digital clock*. The crystal oscillator runs at a high frequency, at a number of Hz which must be an exact power of 2. The divider chain then divides this frequency down to one pulse per second, one pulse per minute and one pulse per hour. At every divider stage, decoders convert the BCD numbers representing hours, minutes and seconds into voltages which operate the correct segments of a six figure set of seven-segment displays.

The diagram, following convention, shows the hours, minutes and seconds displays in the opposite order.

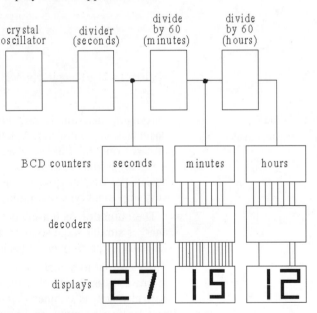

Figure 14.3   Block diagram of a digital clock

If a gate were to be inserted between the oscillator and the first divider, with a reset so that all the displays could be set to zero, the block diagram of Figure 14.3 would be converted into that of a stopwatch.

Registers

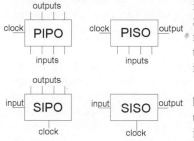

A register is formed from a set of bistables connected together, and such registers are available in IC form. Registers are of four types, because the inputs can be serial or parallel, and the outputs can also be serial or parallel. Serial means working with one terminal, using pulses in sequence. Parallel means that a set of inputs or outputs will be used, with signals on all of them. We class registers as PIPO, SIPO, PISO and SISO, with I meaning input, O meaning output, S meaning serial and P meaning parallel.

The PIPO register is used as a store for binary digits (bits), typically with bits at the inputs read into the register on one clock pulse, and fed out from the outputs on the next clock pulse. The PISO register will accept a set of inputs and feed these bits out from the serial output, one at a time, on each clock pulse. The SIPO register will accept an input serial bit at each clock pulse until the full set of bits is available at the outputs. The PISO and SIPO registers are extensively used in serial communications systems. The SISO register will store an input bit and release a stored bit at each clock pulse, and its main application is as a bit delay.

A-D and D-A converters

An A-D converter (ADC) is a circuit that will convert a varying voltage, such as an audio signal or a measurement of a varying voltage, into a set of digital signals. This is done by sampling the voltage of the input signal at intervals, and converting each value of voltage level into a binary number. This action is an essential part of a digital voltmeter and is also used in recording sound on a compact disc.

The clock rate that is needed is set by the number of samples that need to be taken each second, and also by the number of bits that are needed to represent the size. The most difficult ADC task is in converting sound for CD use. The sampling rate is around 44 thousand samples per second, and 16 bit numbers are used. Instruments like the digital voltmeter can use much slower sampling rates and a smaller number of bits.

The D-A converter (DAC) performs the opposite conversion, from a set of binary numbers into a voltage level. A fast DAC is an essential part of a CD player.

Answers to questions

14.1   The transistors have very little effect because they conduct only for brief intervals. Some transistors, however, might be incapable of switching rapidly enough.

14.2   For a 12-hour clock the first figure is either 0 or 1.

15 Digital inputs and outputs

Switch inputs

Digital inputs can be obtained from A–D converters when an analogue signal is available, but digital inputs can also be obtained from switches. Remember that the keyboard of a computer is a set of switches.

The problem of obtaining digital signals from switches is that a switch is a mechanical component. When a switch is closed, the contacts should ideally come together and stay together, but in fact they usually bounce. Each bounce will generate another pulse, so that the output can consist of a set of pulses each time the switch is closed.

Switch de-bouncing circuits are used to ensure that only one pulse is generated by a switch being closed. There is no problem with bounce occurring when a switch is opened, but for a changeover switch, bounce can occur on either direction of switching. The simplest debouncing method is to connect each input as a CR circuit so that the pin voltages at the input of a gate cannot change quickly. Though this can be effective, it is by no means foolproof, so that more elaborate methods are normally used.

One method of debouncing switch contacts is to connect the switch to a bistable formed from NAND gates. This type of bistable circuit is called an R–S flip-flop, and the circuit, left, shows a typical switch debounce connection. To understand what happen, we need to know how the R–S flip-flop works, and this is summarized in the table, called a *sequence table*. The point to note is that the output changes only when one input is at logic 1 and the other is at logic 0. When the input are both at logic 1, the output remains unchanged at whatever it was last set to, 0 or 1.

When the switch is at position (a), the input to the flip-flop is R = 0, S = 1, and this gives Q = 1 as the output. Now while the switch is being changed over, both inputs will momentarily be at logic 1, but this does not change the output. When the switch contacts are in position (b), the inputs to the flip-flop are R = 1, S = 0, giving Q = 0. If the switch contacts bounce, making R = 1, S = 1, the output remains at 0, so that the switch bounce has no effect.

Another method of eliminating switch bounce (requiring more design effort) is illustrated in Figure 15.1. This uses a CR circuit that does not allow the output to rise quickly when the switch is opened, and this type of action is even more effective when it is combined with the Schmitt circuit.

The Schmitt circuit has the type of characteristic illustrated in Figure 15.1. A small change of voltage at the input has no effect, but larger a change will make the circuit flip over very rapidly. This is a form of hysteresis (see Chapter 6) which does not involve magnetization. When the input voltage decreases, there is no change at the output until the input voltage is at a low level. When this is combined with a CR time constant, switch de-bouncing can be carried out very effectively. When the switch closes, the input and output of the buffer change over. If the switch contacts bounce, the change of voltage at the input is not rapid, because of

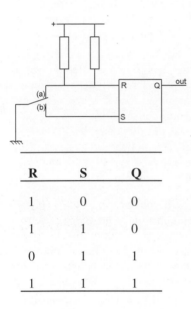

R	S	Q
1	0	0
1	1	0
0	1	1
1	1	1

the time constant, so that the output voltage is unaffected. Buffers and logic gates are obtainable with Schmitt switching characteristics.

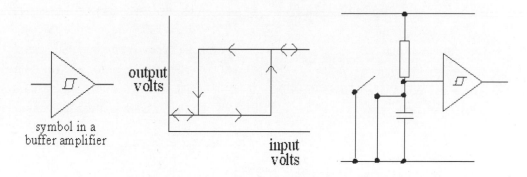

Figure 15.1 Schmitt trigger characteristic and switch de-bouncing circuit

De-bouncing for computer keyboards is carried out by software instructions to the computer processor. When a switch is closed, a time delay starts, and changes in the switch setting have no further effect until the end of the time interval. This interval can be several milliseconds, since you cannot type fast enough to operate a key more than once in that time.

Activity 15.1

Assemble a debouncing circuit (either Schmitt or R–S flip-flop) on a circuit board. Use a push-button switch as a pulse source, and test with a counter to detect bounce. Add the debouncing circuit and verify that the counter no longer detects pulses caused by bouncing.

Output indicators and displays

LED indicators are diodes, see Chapter 8, that emit light when current passes in the forward direction. By using different materials, LEDs can be made to emit light of different colours. Single LEDs can be used to indicate binary numbers by using one LED for each power of two.

LEDs are more commonly used for indicating denary numbers, using ICs that convert binary to BCD, and BCD into LED segments on a seven-segment type of display, Figure 15.2. The amount of current passing needs to be controlled by using a resistor in series with the LED. The brightness of the output is roughly proportional to the current.

The format of LED indicators can be common-anode or common-cathode. For a seven-segment display, for example, each bar can be the cathode of a diode, all of whose anodes are connected, or each bar can be an anode of a diode, with all the cathodes connected. On a common-cathode display, the cathode will be earthed and each anode connection driven to a positive voltage so as to illuminate a bar. On a common-anode display, the anode connection will be made to a positive voltage, and the cathode connections earthed by the driver circuits as required.

The advantages of LED displays are:

- Low voltage operation
- Clear bright display
- Controllable brightness
- Bright and clear when viewed off-axis
- Operate on d.c.

The disadvantages of LED displays are:

- Each LED unit requires a substantial current
- High power consumption
- The colour choices are very limited
- The diodes are destroyed by a reverse voltage
- They are not easily visible in bright light

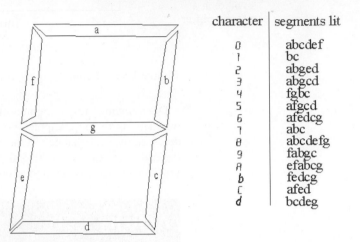

character	segments lit
0	abcdef
1	bc
2	abged
3	abgcd
4	fgbc
5	afgcd
6	afedcg
7	abc
8	abcdefg
9	fabgc
A	efabcg
b	fedcg
C	afed
d	bcdeg

Figure 15.2   A seven-segment display arrangement

LCD displays

LCD displays make use of a quite different principle. They depend on a material that polarizes light when an electrical signal is applied. By sandwiching this *liquid-crystal* material between metal electrodes (one of which is transparent) and passing light to it through a polarizing filter, the light can be switched on or off by altering the voltage across the liquid crystal. The amount of power that is needed is very small, and the brightness of the display is just the brightness of the light that is used.

The voltage supply must be a.c. at a high audio frequency, with no trace of d.c. LCD displays are normally used as reflectors, so that they are most visible in bright light, but backlighting can be used if the displays have to be read in the dark. One severe disadvantage is that the display is clearly visible only when you look straight at it. Visibility is much reduced when you view at an angle.

Both LED and LCD displays can be made in a variety of patterns, of which the seven-segment display is the most common. Another useful type of display is the dot-matrix, using a set of dots for each character. This is used more with LCD than with LED nowadays, and the drawing shows a typical 5 × 7 pattern. Any character of the alphabet or number digit (alpha-

5 x 7 matrix   example of display

numerical character) can be displayed, and using larger matrices such as 8 × 15 allows for foreign characters with accents and oriental characters also to be displayed.

The advantages of LCD displays are:

- They are easy to read in bright light
- The power consumption is very small
- Colour displays are possible
- Very small elements can be made
- Complicated patterns can be manufactured

The disadvantages are:

- A backlight is needed for reading in darkness
- The drive supply must be high-frequency a.c.
- The acceptable viewing angle is small
- Control of brightness is difficult
- D.c. must not be applied

Other formats for these displays are starburst patterns, and bargraphs. Bargraph displays are very commonly used as audio level controls in tape recorders and graphics equalizers.

All forms of display require driver circuits that will accept normal logic level inputs and provide outputs that are suitable for operating the display segments. For number or alphabetical displays, combined decoder drivers are used to implement the logic actions for the signals in Figure 15.2. The decoder/driver and display is often combined so that no wiring is needed other than power supplies and inputs.

Question 15.1

What form of display do you think would be most suitable for a laptop (portable) computer? Give reasons.

Colour LCDs

The new types of colour LCDs exist in three forms. The transmissive displays (TMC) are used with a white back light, and the liquid crystal dots will transmit whatever colour is selected from the whole spectrum of white light. These displays are ideal for use in low-lighting conditions.

Reflective displays (RFC) are used illuminated by white light, and will reflect whatever colour is selected from each liquid-crystal dot. No back light is needed, so that power requirements are very low, but they can be used only in bright lighting conditions, and the perceived colours will alter if the illumination is not white light.

Transflective displays (TFC) combine both TMC and RFC principles, and use both a backlight and the ambient light to provide a display that can be read equally easily in the dark or in sunlight. The user can save power by having the back light off in high ambient light.

The most common back lighting methods use electroluminescent (EL) lamps, light emitting diode (LED) arrays or cold-cathode fluorescent (CFL) tubes. EL back lights are slim, require low power, but have relatively limited life. LED back lights can provide high brightness, uniform appearance, long life, and low cost. CFL back lights feature very uniform brightness and appearance, long life and low power consumption and are the preferred option for transflective displays.

Answers to question

15.1   Transflective CFL, which can be used either in daylight or in darkness, but with low power requirements.

Unit 6

1. Demonstrate an understanding of home entertainment systems and apply this knowledge safely in a practical situation.
2. Demonstrate an understanding of TV receivers and apply this knowledge safely in a practical situation.

16 Audio, radio transmission and reception

Radio waves and modulation

One of the earliest applications of electronics was in the transmission and reception of radio waves. Any a.c. signal passing along a wire is also radiated into the air from the wire. Waves of very low frequency, such as the 50 Hz used for power supplies, will radiate effectively only if the wire carrying them is extremely long (in the region of several *million* metres). However, overhead power lines can readily collect and then re-radiate higher frequencies as interference. Wave energy of higher radio frequencies, (from about 16 kHz upwards) can radiate more easily. This figure should not be confused with the upper audible frequencies because the radiation mechanisms differ. Audio waves propagate via the movement of air molecules that impinge on the ear drums, whilst radio propagation relies on the movement of charged electrons that are too small to influence the human hearing mechanism.

Early in the history of radio, we found that if the radiating wire, called an *aerial* or antenna (see Figure 16.1), would radiate very efficiently if it were cut to a length exactly one quarter of the wavelength of the signal. This is referred to as a *tuned aerial*. Such a *quarter wavelength aerial* behaves to a signal wave as if it were a special kind of resistor, one which dissipates the energy of the signal in the form of radio waves rather than as heat. This *electromagnetic energy* has two components, (electrostatic and electromagnetic) which act mutually at right angles to each other and also at right angle to the direction of propagation. Tuned aerials will be further considered in Chapter 17.

The way the waves are transmitted depends on how the wire is arranged. If the wire is vertical, the waves are vertically polarized, meaning that the electrostatic part of the wave is vertical. For a horizontal wire, the waves are horizontally polarized, with the electrostatic part of the wave horizontal. The importance of this is that a receiving aerial should be similarly polarized if it is to work at maximum efficiency.

Aerials are also used to receive radio waves, converting the waves travelling through the air into signals of voltage or current in a wire leading from the aerial. Again, the exact length of the receiving aerial determines the efficiency; but in any case only a tiny fraction of the power radiated by a transmitter is ever picked up by the aerial of any one receiver.

Aerials for both transmission and reception can take on a wide variety of forms, but those used for the reception of radio broadcast signals may either be untuned or tuned devices. These latter are sometimes described as *resonant* aerials. Generally for portable operation of long and medium sound broadcasts either a telescopic rod or ferrite rod device is used. For the VHF sound broadcasts where higher quality is expected the aerial is commonly of the resonant type. The telescopic rod acts somewhat as a short (less than quarter wavelength) monopole, whilst the ferrite slab tends to concentrate the electromagnetic component of radiation within itself to induce a higher signal level in receiver tuned input circuits.

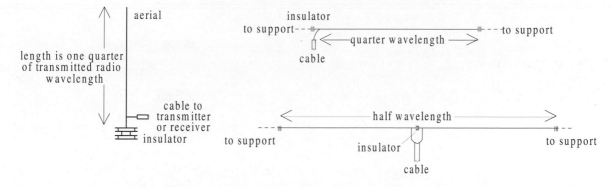

Figure 16.1   Aerials: vertical and horizontal polarization

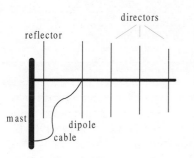

The efficiency of an aerial can be improved by using a half-wave dipole (see Figure 16.1) and then adding unconnected wires or rods in parallel, as illustrated, left. These increase the *gain* of the aerial meaning the voltage output for a given strength of radiated signal. They also make the aerial more directional so that it picks up more of the wanted signals and less of any unwanted signals from other directions. The rods or wires that are placed between the half-wave aerial and the transmitter are called *directors*. Rods or wires that are placed behind the half-wave section are called *reflectors*, and the whole aerial is called a Yagi.

The wire and rod aerials detect the electrostatic component, whilst ferrite rods by comparison rely on the electromagnetic component of the radiated wave energy, using the fact that ferrite is a soft magnetic material that concentrates the magnetic lines of flux, so providing some gain from an electromagnetic wave. A coil wound round the ferrite will convert the changing magnetic flux into a varying voltage at the frequency of the signal waves.

A transmitted signal which consists wholly of a steady radio frequency is called a continuous wave (CW) transmission. Such a transmission is useless except as a way of establishing that a transmitter is operating. To convey a signal, some means of varying the wave must be found.

Question 16.1

What is the length of a quarter-wave aerial for 850 MHz?

Modulation

The oldest known method of carrying information by radio waves is to switch the radio wave on and off in a pattern of longer *dashes* and short *dots*. This is called the Morse code, and it was used by Samuel Morse for telegraph signalling in 1838. A carrier interrupted in this way whether by Morse, or by any other of the many different types of telegraph codes, is called an interrupted continuous wave (ICW) transmission (see Figure 16.2).

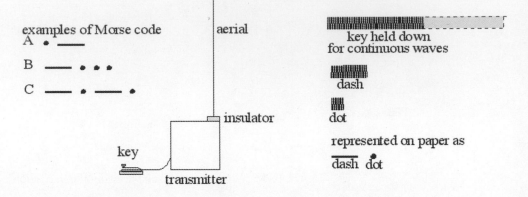

Figure 16.2   ICW transmission using the Morse code

Though Morse code has been extensively used since the early days of radio for emergency transmissions and by radio amateurs, it is a relatively slow method of communication since every letter of the message has to be coded and transmitted separately. Since the early 1920s, therefore, radio communication has concentrated on varying either the amplitude or the frequency of the carrier wave itself so as to transmit audio and other signals of lower frequency. Such a process is called *modulation* and we say that the carrier wave is *modulated* by the lower frequency (AF) signal.

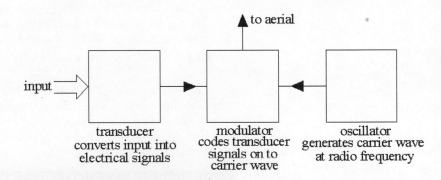

Figure 16.3   How a transmitter works

High frequency carriers can carry many different kinds of information such as audio signals at sound frequencies, video signals from a TV camera, or other waveforms such as digital signals from computer to computer, the transmission of facsimile copies of documents, or radar. The principle of operation is shown in Figure 16.3.

The waveforms that provide the information are derived from *transducer* devices (see Chapter 11) which convert one form of energy into another. The transducers that are of interest in electronics are those that use or provide electrical signals. The transducer for sound is the *microphone* and the transducer for pictures is the TV camera *image sensor*.

Transducers at the receiving end, Figure 16.4, of a radio system will convert the electrical signals back to the original type of signal. For sound signals, this transducer is the *loudspeaker*; for TV signals it is a *cathode-ray tube* (CRT). The aerials of both transmitter and receiver are also transducers, either converting electrical signals carried on a wire into electromagnetic waves, or vice versa.

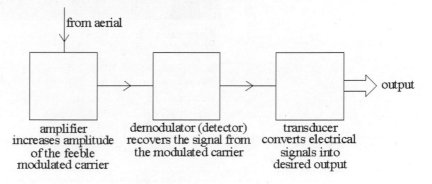

amplifier increases amplitude of the feeble modulated carrier

demodulator (detector) recovers the signal from the modulated carrier

transducer converts electrical signals into desired output

Figure 16.4   How a receiver works

Amplitude modulation

One method of carrying an audio or video signal using radio waves is called amplitude modulation (AM) see Figure 16.5.

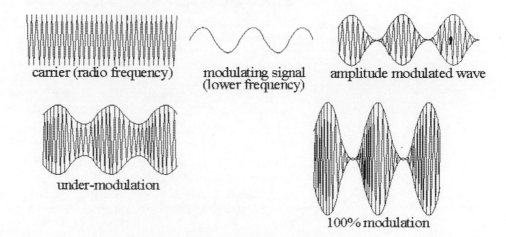

carrier (radio frequency)

modulating signal (lower frequency)

amplitude modulated wave

under-modulation

100% modulation

Figure 16.5   The amplitude modulated carrier

An amplitude modulator is a circuit into which two signals are fed: the carrier wave at a high (radio) frequency and the modulating signal at a lower (audio or video) frequency. The output of the modulator is a signal at carrier frequency whose amplitude exactly follows the amplitude changes of the modulating signal. The greater the amplitude of the modulating signal, the greater the *depth of modulation* of the carrier. Excessive amplitude of the modulating signal causes *over-modulation* of the carrier itself, resulting in distortion when the signal is recovered by demodulation.

Activity 16.1

From the Colpitts oscillator circuit built for the activity in Chapter 12 remove the 0.1 μF capacitor across the 22K resistor. Connect the low resistance output terminals of an audio signal generator to the base of the transistor, and to earth making the connection to the base of the transistor through a 1 μF capacitor. Connect the oscilloscope to the output of the oscillator, switch on the supply (but not the audio signal generator), and observe the sine wave on the oscilloscope screen.

Now set the signal generator to 400 Hz, minimum amplitude, and switch on. Observe the modulated signal as the amplitude of the audio signal is increased. Sketch the shape of the trace at several different amplitudes of modulation.

Question 16.2

Overmodulation causes distortion, but why is it also important to avoid undermodulation?

Frequency modulation

A second important type of modulation is called *frequency modulation* (FM). It uses the amplitude of the modulating signal to alter the frequency of the carrier, but the amplitude of the carrier itself remains constant.

The alteration of frequency caused by a modulating signal is called the *frequency deviation*. In *wideband FM* systems deviations of 75 kHz or more are produced; in *narrow-band FM* systems deviations are limited to a few kHz only. The principle is shown in Figure 16.6.

At one time, all radio broadcasting used AM, and the FM system, invented by Major Edwin Armstrong (who also invented the superhet receiver principle) was hardly used. The need for high-quality broadcasting, particularly in stereo, and the number of stations using the AM system on medium-wave frequencies, led to the adoption of FM from the 1960s onward.

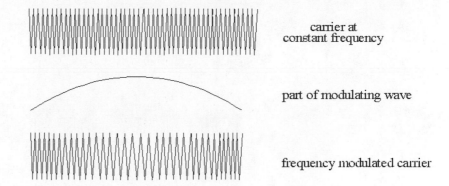

Figure 16.6 The principle of frequency modulation, showing one quarter of a modulating wave affecting the frequency of the carrier

Pulse modulation systems

Another system called *pulse modulation* operates by switching the carrier on and off in a pattern resembling Morse code, but at a very much faster rate. Pulse modulation is used mainly for radar signals, or in communications systems using carriers at frequencies above 1000 MHz. The basic principle is illustrated in Figure 16.7.

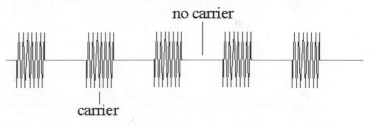

Figure 16.6 The principle of simple pulse modulation. The carrier is at constant frequency and amplitude during a pulse

A pulse signal can be modulated for communications purposes in several different ways. *Pulse amplitude modulation* (PAM) systems alter the amplitude of the pulses in the same way as the amplitude of a continuous carrier is modulated in the AM system. Note, however, that each pulse consists of a large number of cycles of a carrier at a very high frequency, and that the pulse repetition rate (p.r.r.) needs to be greater than the modulating frequency.

Pulse width modulation (PWM) systems use the modulating signal to change the width of the pulse of carrier waves. A sine wave modulation, for example, appears as illustrated in Figure 16.8, where the widest pulse of the modulated carrier corresponds to the positive peak of the signal and the narrowest pulse to the negative peak. The amplitude and frequency of the carrier waves remains constant, only the number of cycles of carrier in each pulse is changed.

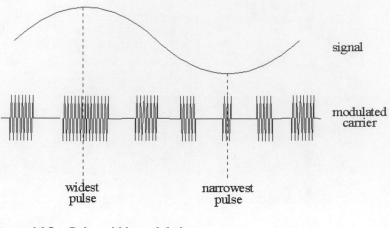

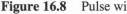

Figure 16.8   Pulse width modulation

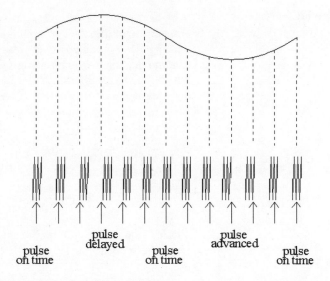

Figure 16.9   Pulse position modulation

Pulse position modulation (PPM) leaves both the amplitude and the width of the pulses unchanged, but alters the time intervals at which they occur. A pulse repeated at regular intervals carries zero modulation, but if the timing of a pulse is varied so that it occurs either before or after the 'normal' time of its recurrence, this variation from its set time is used to carry information of the amplitude of a signal. When the arrival of a pulse is delayed after its set time, it indicates positive peak of signal. When it arrives before its set time, it indicates negative peak of signal. The system is illustrated in Figure 16.9, in which arrows indicate the set, or 'normal', times of arrival of the pulses.

Pulse code modulation (PCM) is a very popular technique for modern communications which converts the analogue signal to be transmitted into a digital format by the process of sampling or measuring the amplitude at precise intervals of time. The amplitude of each sample is then converted into a binary code and this is then transmitted by any one of the forms of carrier modulation. Because PCM and PPM systems handle constant amplitude and width pulses, both are relatively immune to interference problems. PCM is used for digital recording systems (such as CD) as well as for digital transmissions like digital TV and Nicam sound.

- In many applications, such as digital TV, the PCM signal is modulated on to a conventional RF carrier.

Multiplexing

The electromagnetic spectrum is not infinite, and some parts cannot be used at present because we have no simple way of generating, modulating or detecting the signals. Various methods can be used to maximize spectrum availability and *multiplexing* is a way of achieving this. Signal multiplexing means that two or more communications channels are using the same medium simultaneously. This, for example, allows digital TV to use more channels without using a greater total bandwidth.

Time multiplexing is a system that allows the same carrier frequency to be used many times over by allocating that carrier to each specific channel in turn for brief periods of time. Signals for any specific transmitter are interleaved with information from other transmitters. The same carrier can be used at least twice for analogue modulation, by allocating differently phased versions of the carrier to different modulating (baseband) signals. Digital and pulse modulation systems adapt very easily to some types of multiplexing to save bandwidth. However, all these systems require very careful synchronization at the receiver in order to recover the original modulating signals.

The modulated signal

The simplest possible unmodulated carrier is a sine wave which has, of course, a single value of frequency, and carries no useful information. Whenever such a carrier is changed in any way, however, other frequencies can be detected in it. These new frequencies that are caused by modulation are called *sidebands*. Any serviceable receiver must be able to receive these sidebands as well as the carrier wave itself. This is because the sidebands, not the carrier, contain the desired information.

An amplitude modulated (AM) carrier has the simplest sideband structure. Imagine a sine wave carrier at 400 kHz modulated by an audio signal which is a 2 kHz sine wave. The effect of modulation is to produce two new sideband frequencies, one at 402 kHz (carrier plus modulation frequency) and the other at 398 kHz (carrier minus modulation frequency), in addition to the carrier frequency of 400 kHz itself. When the modulation is at its maximum possible, the amplitude of each sideband should be exactly half the amplitude of the unmodulated carrier. The effect is illustrated (in exaggerated form) in Figure 16.10.

- **Note** that the spacing between the carrier and each sideband is equal to the modulating frequency

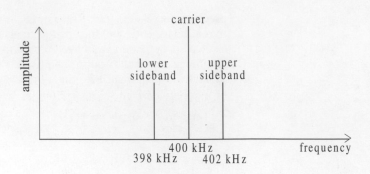

Figure 16.10 Sidebands of a carrier modulated by a 2 kHz sine wave

Over-modulation occurs when the modulating signal has a peak-to-peak amplitude greater than half the peak-to-peak amplitude of the carrier (see Figure 16.11). Over-modulation causes severe distortion of the waveform of the modulating signal and is one of the reasons why AM is no longer used to transmit high quality sound signals. At the transmitter, over-modulation is prevented by means of limiting circuits which reduce the extent of the distortion.

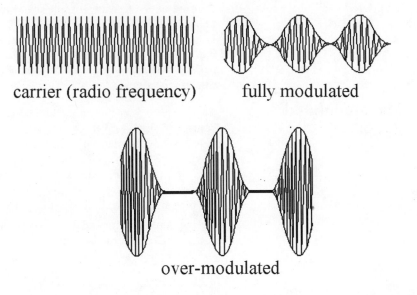

Figure 16.10 Effect of over-modulation

Under-modulation causes no distortion, but it makes the received audio signal very faint and therefore easily drowned out by electrical noise and interference signals. Persistent over or under-modulation is a clear sign of a faulty modulator system.

When a carrier is modulated by an audio signal of speech or music, the audio signal that results is not a single sine wave, but a mixture of frequencies. The range of frequencies present in the modulation is called the *bandwidth* of the modulation or the baseband signal. Speech can be satisfactorily transmitted over a bandwidth of a mere 3 kHz, but music requires bandwidths up to 20 kHz for high quality signals to be satisfactorily received and reproduced.

A carrier which has been amplitude modulated by audio or video signals therefore contains sidebands which themselves consist of a mixture of frequencies and extend from $f_c - f_m$ to $f_c + f_m$, where f_c is the carrier frequency and f_m is the highest modulating frequency.

Bandwidth of transmission is (854 – 846) kHz = 8 kHz, which is twice the highest modulating frequency

The upper sideband contains all the frequencies between f_c and $f_c + f_m$, whilst the lower sideband contains all the frequencies from $f_c - f_m$ to f_c. Given, for example, a carrier frequency of 850 kHz and modulation frequencies extending to 4 kHz on either side of it, the upper sideband would contain all the frequencies from 850 kHz to 854 kHz, and the lower sideband all the frequencies from 850 kHz to 846 kHz. Figure 16.12 shows this sideband structure which should normally be a mirror image about f_c (850 kHz).

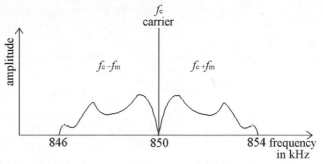

Figure 16.12 Sidebands of a carrier amplitude-modulated by a typical audio wave

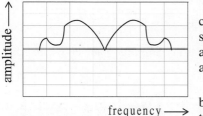

A graph that plots the amplitude of signals against a scale of frequency is called a *spectrum*, and an instrument which displays such a graph for any signal is called a *spectrum analszer*. The drawing, left, shows the typical appearance of the trace of a typical AM signal displayed on a spectrum analyser.

When two AM transmitters that are in the same geographical area are broadcasting on the same frequency, it will be impossible to receive either transmission clearly because of interference from the other. Even if their broadcasts are on different frequencies, interference will also be caused if the sidebands of the two modulated signals overlap. To avoid interference, the carrier frequencies must be separated by *at least* twice the maximum frequency of the modulating signals, Figure 16.13.

Unfortunately, in many parts of the developed world, including Western Europe, there are many transmitters operating on the medium wave frequencies. Reasonable reception is made possible, partly by limiting the bandwidth of modulation to 4.5 kHz (which makes the transmission of high quality sound impossible), and partly by international agreement on the frequency and power output of individual transmitters. Pirate transmitters do not observe these agreements, and are therefore *always* a cause of interference. Figure 16.14 shows a typical spectrum of overlapping sidebands caused by transmitter frequencies being too close which indicates how adjacent channel interference (ACI) arises.

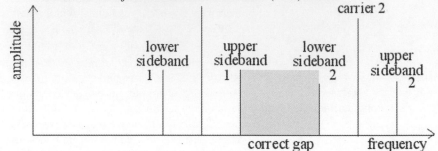

Figure 16.13 Correct separation of carrier waves

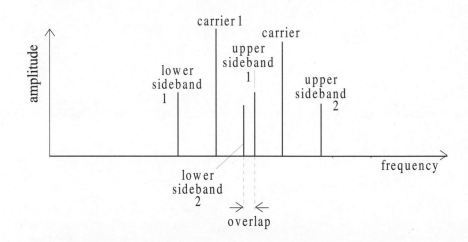

Figure 16.14 The sidebands overlap causing adjacent-channel interference

Question 16.3

A 1.5 MHz carrier is amplitude modulated with a sine wave at 5 kHz. What sideband frequencies are created?

Frequency modulated (FM) signals have a much wider spectrum. FM broadcasts in the UK have an audio bandwidth of 15 kHz, with the modulated bandwidth for each sideband as wide as 200 kHz. The bandwidth of a frequency modulated signal cannot be calculated so easily as that of an amplitude modulated signal. It is always by far the wider of the two for the same audio modulation. There is no room on the medium wave band for FM signals, and these make use of the VHF frequencies approximately between 86 MHz and 114 MHz.

Pulse modulation requires a greater bandwidth than either AM or FM, and it cannot be used at frequencies on which too many transmitters are radiating at the same time. The bandwidths required are so great that carriers of very high frequency, of the order of 1000 MHz (1 GHz) or more may have to be used. At such frequencies as these, pulse modulation is often the only form of modulation which can be applied by the devices which generate the carrier waves.

Demodulators

At the receiver, a *demodulator* circuit is used to extract the desired information signal from the modulated carrier. The recovered signal must be free from any trace of the carrier, and a low-pass filter is an essential part of the circuit.

Because diodes are used in demodulation circuits, it is usually possible to obtain a steady d.c. voltage that is proportional to the average amplitude of the carrier. This can be used to control the gain of the early stages of the receiver (see later in this chapter).

Each different type of modulation system requires its own peculiar demodulation circuit to get optimum results; and very elaborate demodulators are needed for some pulse modulation systems, particularly PCM.

Comparison of different modulation systems

Amplitude modulation (AM) advantages:

- Comparatively simple modulation circuits.

- Very simple demodulation.

- Calculations of bandwidth are easy.

Amplitude modulation (AM) disadvantages:

- Much transmitter power is wasted because only one sideband is needed to carry information, the carrier and the other sideband being unused by the receiver.

- Signals caused by electrical storms or by unsuppressed electrical machines cause interference with the received signals.

The range and efficiency of AM transmissions can be much improved by using *single sideband*, or *suppressed carrier*, very often both (SSB), systems. In these systems, either an arrangement of filters or a different type of modulator circuit at the transmitter is used either to remove one of the carrier sidebands almost completely, or to reduce the amplitude of the carrier, or both. The penalty which has to be paid is a much more complex demodulator circuit. These systems have never been used in general broadcasting but they are commonly used in high-grade communications systems and by radio amateurs.

Frequency modulation advantages:

- Simple modulation circuits.
- Carrier amplitude constant, so that transmitter range can be greater than with AM.
- Freedom from interference because interference signals do not alter the frequency of the transmitted signal.

Frequency modulation disadvantages:

- Complex demodulator circuits.
- Wide bandwidth requirement.

Pulse modulation advantages:

- Simple modulation circuits.
- Often the only type of modulation possible with some oscillators.
- Multiplex transmission made possible.

Pulse modulation disadvantages:

- Very wide bandwidth needed, so that pulse modulation can only be used with carriers of very high frequency.
- PCM requires elaborate circuits for modulation and demodulation.

Measurement of modulation

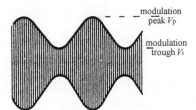

modulation peak V_p

modulation trough V_t

The extent of modulation of an AM transmission is measured by reference to its percentage depth of modulation. Take as an example the amplitude modulated waveform illustrated left. The amplitudes of the modulation peak (V_p) and of the modulation trough (V_t) are measured, and the percentage depth is given by the expression:

$$\frac{V_p - V_t}{V_p + V_t} \times 100\%$$

The value of this expression must never be allowed to exceed 100 per cent if the signal is to be saved from severe distortion, and in practice it should not even approach 100 per cent.

The concept of depth of modulation for FM is rather complex and is based on the ratio of carrier frequency change to the frequency of the modulating signal that produced the change. The term *deviation* is used to describe any change in the carrier frequency from its mean or unmodulated value. The *modulation index* is defined by

$$\frac{\text{carrier frequency deviation}}{\text{modulating frequency}}$$

and this value varies considerably. However a limiting value is specified for any FM system in the form of the *deviation ratio* which is given by:

$$\frac{\text{maximum carrier frequency change allowed}}{\text{maximum modulating frequency allowed}}$$

For the European VHF radio service the absolute amount of deviation has been standardized at ±75 kHz.

In all FM broadcasts, the modulation index has a value considerably in excess of unity. By contrast with an AM transmission, however, no sudden change of waveshape can be caused by over-modulation, whereas with AM, a single modulating frequency generates only a single pair of side frequencies in the complex wave. With FM each single modulating frequency generates many sideband pairs. The FM bandwidth is constant, irrespective of the modulation frequency or amplitude, so that a low modulating frequency gives rise to a larger number of sideband pairs than does a higher modulating frequency.

Transmitter and receiver block diagrams

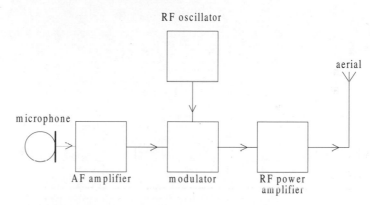

Figure 16.15 Block diagram of AM transmitter

Figure 16.15 shows the block diagram of an AM transmitter. The audio signal originates at the transducer, a microphone, which converts sound waves into electrical signals. The AF amplifier amplifies these feeble signals (they have an amplitude of one millivolt or less) to the amplitude required to modulate the carrier wave.

At the same time, an oscillator tuned to a high radio frequency generates a carrier wave. This oscillator stage may be followed by several other stages of frequency multiplication and amplification, but eventually carrier wave and audio signal are combined in the modulator which produces at its output an amplitude-modulated RF signal.

This modulated RF signal is amplified in the power amplifier (PA) to produce a signal which can supply the aerial (the output transducer of the transmitter) with an alternating voltage and a large alternating current. The PA stage is required because the aerial consumes much power, dissipating it in the form of radio waves of electromagnetic radiation. This power, typically anything from a few hundred watts to several hundred kilowatts, cannot be supplied from an oscillator stage directly, because the loading of the aerial greatly reduces the stability of the oscillator. Many AM transmitters, however, avoid the need to amplify a modulated signal by carrying out modulation at the PA stage.

Apart from the audio amplifier, every amplifying stage of the transmitter is tuned to operate at the frequency of transmission. Filter circuits ensure that the signal fed to the aerial contains only the desired output carrier and its sidebands.

The FM transmitter

For the FM transmitter, Figure 16.16 shows that the audio signal is still used to modulate the carrier, as in the AM transmitter, but now in such a way as to cause changes in its frequency rather than in its amplitude. The frequency modulated wave is then amplified in the power stages before being radiated by the aerial system. Table 16.1 offers a convenient summary of the detailed effects of both types of modulation. The methods used for digital radio transmissions will be looked at later.

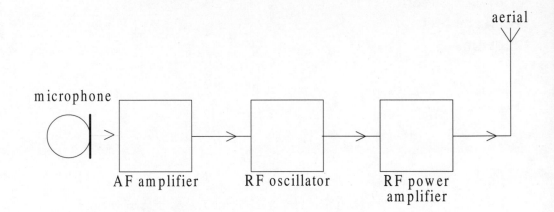

Figure 16.16 Block diagram of FM transmitter

Table 16.1 Summary of effects of modulation

	AM	*FM*
Amplitude of modulating signal	Varies carrier *amplitude*	Varies carrier *frequency*
Frequency of modulating signal	Controls rate of change of *amplitude*	Controls rate of change of *frequency*
System constant	Carrier *frequency*	Carrier *amplitude*

Question 16.4

What, apart from simplified circuitry, is the attraction of carrying out modulation at the PA stage of an AM transmitter?

Receivers

Signal to noise ratio

In any transmission channel, noise and interference can be considered as a killer of information. It is therefore most important that any receiver should not add too much noise to a signal input that may already have been degraded. Interference can be created by other communications channels that occupy the same or adjacent frequencies. Noise is mostly created from other electrical systems, power lines and even the ionosphere. The annoyance factor or quality of the processed signal can be expressed in terms of the system *Signal to Noise Ratio* (S/N). This factor is expressed as; 10 log (Signal power/Noise power) dB or 20 log (Signal voltage/Noise voltage) dB.

Superhet receiver

Figure 16.17 shows the block diagram of an AM receiver typical of a transistor radio intended for medium or long-wave reception. Like practically all radio receivers, its operation is based on the *superheterodyne* (superhet) principle.

In a superhet receiver the incoming signal, whatever its frequency, has that frequency converted to one value of intermediate frequency (IF), which is lower than the radio frequency. This conversion makes amplification easier, for it is simpler to design a high gain amplifier that is tuned to a set frequency than one which needs to have its tuning altered to receive another transmission. A superhet receiver carries only a few variable tuned circuits, because most of the amplification is carried in the IF amplifier that used fixed tuning.

In Figure 16.17, the stage labelled *RF filter* selects the wanted signal from all the other signals picked up by the aerial. This filter therefore must have variable tuning. It acts also to reduce the amplitude of any oscillator frequency which might otherwise be re-radiated from the aerial.

The *oscillator* circuit (or *local oscillator*) also is variably tuned, but in such a way that its frequency is always 465 kHz higher than that of the incoming signal. Preset capacitors called *trimmers* and *padders* maintain this frequency difference, which is the IF frequency, over the whole of the tuned range. The oscillator is said to be *tracking the input* correctly when the frequency difference, or IF, remains correct over the full range.

The incoming signal and the oscillator sine waves are mixed in the *mixer* stage, producing two new frequencies which are also modulated. One of these new modulated frequencies is the IF, which is in this example 465 kHz the difference between the frequencies of the oscillator and of the incoming signal. The second modulated frequency produced in the mixer stage is a signal whose frequency is equal to the sum of the frequencies of the oscillator and of the incoming signal, known as the *image frequency*.

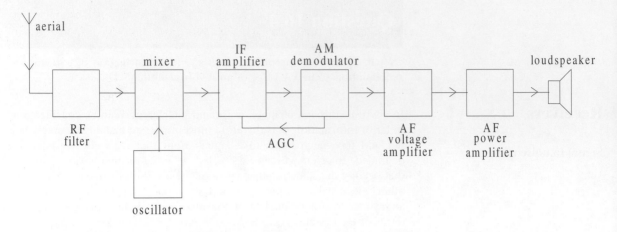

Figure 16.17 Block diagram of AM receiver

The mixer also acts as a tuned amplifier that is tuned to the frequency of the IF. For this reason, only the IF output is allowed to pass to the IF amplifier. All the other frequencies of incoming signal, oscillator sine wave and the sum frequency are blocked or rejected. The IF amplifier, in its turn, greatly increases the amplitude of the IF signal, which itself now carries the same modulation as did the incoming signal.

The modulated and amplified IF signal is applied to the AM *demodulator*, which produces two outputs. One is the modulating signal itself, which is at audio frequency and free of any trace of the intermediate frequency; the other is a d.c. voltage which is always proportional to the average amplitude of the IF signal. When a weak or distant transmitter is being received, the incoming signal will produce at the mixer an IF signal of very low amplitude.

Even after amplification in the mixer and IF amplifier stages, this amplitude may still be low. On the other hand, a nearby or powerful transmitter might provide a signal that directly breaks through into the IF amplifier stages. This then causes interference or overloading. Both lead to distortion.

The d.c. that is produced (by the rectifying effect of the diode) in the AM demodulator is used to minimize the effects of these two extremes, by being fed back to control the gain of the IF amplifier itself. What happens is that a small amplitude signal at the demodulator gives rise to a small d.c. feedback signal which permits the IF amplifier to operate at full gain.

A large amplitude signal at the demodulator gives rise to a large d.c. signal, which is fed as *negative bias* to the IF amplifier and causes it to operate at much reduced gain. By the use of this automatic gain control (*AGC*) circuit, the signal at the demodulator is kept to an almost constant level even when there are great variations in the amplitudes of incoming signals.

The AF signal from the demodulator passes to an audio *frequency voltage amplifier* incorporating a *volume control* which increases or decreases AF gain. This amplifier sometimes includes also a *tone control*, which increases or reduces the gain at low and high frequencies respectively. The AF signal is then fed to a *power amplifier*, which boosts it sufficiently to drive the output transducer, a loudspeaker.

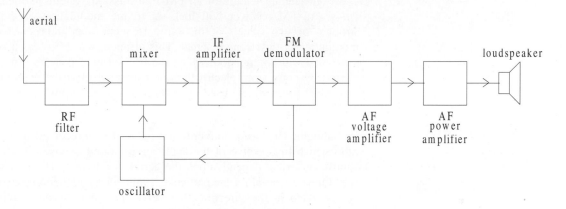

Figure 16.18 Block diagram of the FM receiver

Figure 16.18 shows an FM receiver whose working principles are similar. The superheterodyne principle is again used, but because of the much higher incoming frequency range (90–108 MHz) and the greater bandwidth (about 200 kHz), an IF of 10.7 MHz is required.

Another difference is that a tuned *RF amplifier* is placed in series between the aerial input and the mixer stage. This has two effects:

(a) To raise the amplitude of the desired RF signal well above the amplitude of unwanted interference.

(b) To prevent the oscillator frequency leaking back through stray capacitances from the mixer to the aerial, from which it could cause severe interference for other receivers.

The mixer, oscillator and IF stages act in much the same way as do their equivalents in a medium wave AM receiver. The demodulator, however, is a special FM type, usually called a *discriminator*, because AM detectors cannot in normal operation produce an output from an FM signal. In any case, the circuit must produce no output when the amplitude of the IF signal changes. Only changes in frequency of the wanted signal should result in an output from the demodulator.

The d.c. output from the FM demodulator is of a type different from the d.c. output of the AM demodulator. When the FM demodulator is correctly adjusted, its d.c. output is zero when the IF is centred at the correct frequency which means when the receiver is correctly tuned. The correct tuning of the FM receiver, however, depends on the correct tuning of the oscillator, so this d.c. output is used to control the oscillator frequency.

This arrangement is known as *automatic frequency control* (AFC). It is needed because the frequency of a variable frequency oscillator working in the range 86–110 MHz (or higher) is very easily affected by small changes either of temperature or of supply voltage. The AFC voltage acts promptly to correct these small changes, so keeping the whole receiver tuned to the correct transmission frequency.

An outstanding feature of all FM transmission systems is their *capture effect*. An FM receiver will lock on to the modulation of the slightly stronger of two signals of the same frequency and reproduce only the modulation of the stronger signal. This efficient *selectivity* (as it is called), combined with the freedom which the FM system gives from interference caused by noise from electrical storms or other electrical equipment, has led to FM being widely used in communications equipment, especially for mobile radio.

General installation criteria for receivers

Probably the most important criterion relates to the received signal to noise ratio which irrespective of the application (sound or vision systems) to a significant extent depends upon the aerial installation. The first step is to select the best aerial for the particular application and then make sure that it is orientated in the direction to receive maximum wanted signal level, consistent with the minimum of interference from other signals in the same waveband.

Subcarriers

More than one modulating signal at a time can be transmitted on a single carrier by making use of a *subcarrier* (a form of multiplex operation). A subcarrier is a sine wave whose frequency is greater than the normal frequency range of the modulating signal, but less than the frequency of the main carrier. This subcarrier can be modulated by one, or even two, other signals at a time. The modulated subcarrier and the other modulating signal are then both modulated in turn on to the main carrier, see Figure 16.19.

Both stereo sound broadcasts, which use a 38 kHz subcarrier, and colour signals for TV, which use a 4.43 MHz subcarrier, work on this principle of using subcarriers to pack more information into a modulated signal. The NICAM stereo system that is used for providing stereo sound for TV also uses a subcarrier for the stereo signals, with the mono signal transmitted independently for compatibility with older receivers.

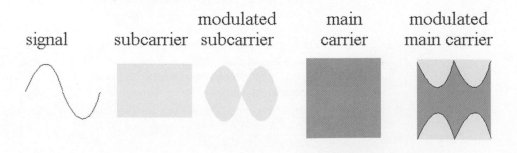

Figure 16.19 Using a subcarrier

The subcarrier is shown amplitude modulated by the signal, and the main carrier is amplitude modulated by the subcarrier in turn. This is a representation only of a process that is almost impossible to illustrate because of the very large differences in frequency. In addition, amplitude modulation is generally used only for the final modulation, or not at all.

Table 16.2 General deployment of the broadcast frequency spectrum

Waveband	Designation	Modulation	Bandwidth	Service	Channel spacing
160–225 kHz	Long wave	AM	9 kHz	Wide area	8 or 9 kHz
525–1605 kHz	Medium wave	AM	9 kHz	Wide area	8 or 9 kHz
88–108 MHz	VHF	FM	200 kHz	Small area	2.2 MHz
470–854 MHz	UHF	AM/FM	5.5 MHz	Television	8 MHz
1.5 or 2.5 GHz	Microwave	Digital	1.5 MHz	DSB	Single channel

- **Radio data system (RDS)**. This is an additional service carried by a number of VHF FM transmitters to provide traffic information for motor transport users. In addition, because many of the higher powered FM transmitters simultaneously radiate the same programme, the receiver can automatically retune to the same programme as the vehicle moves from the service area of one transmitter to another. The low speed data is carried on a 57 kHz subcarrier which is phase modulated at a data rate of about 1.2 kbit/s which occupies a bandwidth of about 4.8 kHz. A similar service known as ARI is in use on mainland Europe but with a different modulation technique.

- **NICAM** (Near instantaneous companded audio multiplex). The normal monophonic sound carrier for UK PAL television is set 6 MHz above the vision carrier. The NICAM stereo signal which is transmitted at a bit rate of 728 kbit/s using a form of phase modulation is located at 6.552 MHz, (9th harmonic of the bit rate). The NICAM signal can carry either a stereo pair or alternate monophonic language channels.

- **Digital audio broadcasting (DAB)**. This service is intended to provide CD quality audio transmissions. It operates in different parts of the world either in the 250/350 MHz (UK) band or 1.4/2.4 GHz, the band used depends upon the local availability of frequency spectrum. The system was developed basically for use in moving road vehicles without introducing multipath distortion effects. Each transmitter radiates a single multiplex of up to 6 programmes using bit rate compression to the MPEG (Motion Picture Expert Group) standard. All transmitters within a large region radiate the same programme multiplex on the same main carrier frequency to form a single frequency network (SFN). Each audio channel can include either 2 separate mono signals in a bi-lingual mode or a stereo pair. The system employs COFDM (Coded Orthogonal Frequency Division Multiplex) transmission techniques with multiple carriers for each programme data stream. The UK system uses 256 carriers separated by 1 kHz for each of the 6 programmes (1536 in total) using a form of phase shift modulation. The multiplex bit rate is about 2 Mbit/s which produces a transmission bandwidth of about 1.5 MHz.

- Whilst the channel spacing used for high powered VHF FM transmission is usually 2.2 MHz, the spacing for low powered stations is often as low as 200 kHz.

- Radio data service (198 kHz long wave transmissions). The high power (500 Kwatt) stations in this group have carrier frequencies that are accurate enough to be considered as frequency standards. They are equipped to broadcast using phase modulation of the carrier, a low speed data service that consists of 16 different channels, one of which is a highly accurate time clock.

- For future service expansion, frequency spectrum has been allocated by the ITU (International telecommunications Union) to accommodate digital radio broadcasting in the medium and long wave (AM) bands. This system which is known as Digital Radio Mondiale (DRM), is designed to provide FM audio quality without the interference problems associated with AM but within the same bandwidth. Like DAB, DRM employs OFDM modulation with 228 separate closely spaced carriers, but with an advanced form of MPEG audio compression.

Tape recording

Tape recording depends on the use of tape heads, see Chapter 6, using a material such as Permalloy with a very small gap cut into the ring, Figure 16.20. The ring has a coil wound round it, well away from the gap, and the tape is made to travel a path which feeds it slowly past the gap.

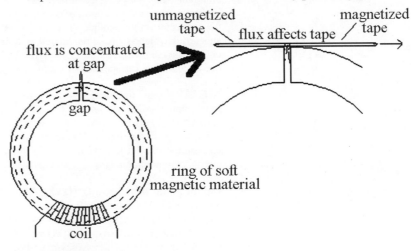

Figure 16.20 How tape is magnetized in the pattern of a signal wave. The tape is moved past the gap in the head, in close contact, and is magnetized by the flux that bulges out of the gap

When a signal current (from an amplifier) flows through the coil, a large alternating magnetic flux is created inside the soft magnetic material. Some of this flux emerges at the gap and the magnetic material that is used as a coating on the tape is magnetized by this flux as it passes across the tapehead gap. Because this coating is a *hard* magnetic material, it retains this magnetism and a permanently magnetized section of tape is produced. Because the tape is moving, each piece of tape has been magnetized by a different part of the incoming signal.

A graph of magnetism plotted against signal strength is not linear because of the hysteresis effect (see also Chapter 6), so a bias signal at about 80 kHz or more, must be added to the audio. The effect of this is illustrated in Figure 16.21. The bias allows the slower-changing audio signal to make use of a portion of the graph that is more linear. The bias signal is obtained from the erase oscillator, using an attenuator. A filter is also connected between the recording amplifier and the recording head to prevent the erase signal reaching the amplifier circuits.

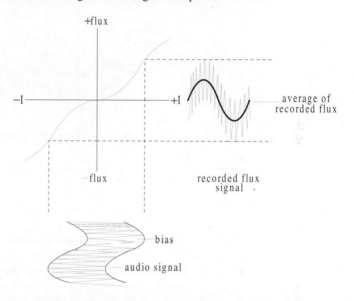

Figure 16.21 The non-linearity of the magnetization graph is overcome by adding a high-frequency bias signal

To replay a recorded tape, the tape is moved, at the same speed as when it was recording, past the same head. The changes in magnetic flux from the tape as it moves past the gap induce a corresponding magnetic flux in the head, which in turn induces voltage signals in the coil wound round the head. These signals, whose amplitude will be around 1 mV, are then amplified in the usual way.

In practice, the signals are deliberately pre-distorted (equalized), both on record and replay, to compensate for imperfections in the tape material and the heads. Tape is erased using a separate erase tape head, with a larger gap, that is fed with a high frequency signal of large amplitude from the oscillator circuit that also provides the bias signal. As the tape moves past the relatively large gap on the erase head, it is magnetized in each direction alternately by succeeding cycles of the erase signal. The amplitude of this signal decreases cycle by cycle as the tape moves away from the centre of the gap, and the action eventually leaves the tape completely de-magnetized.

Cheap tape recorders (including the miniature type used for taking notes) erase by means of a flow of d.c. to the erase head, or even by the use of a permanent magnet. Although this method provides erasure, it leaves behind a rather large noise signal on the tape.

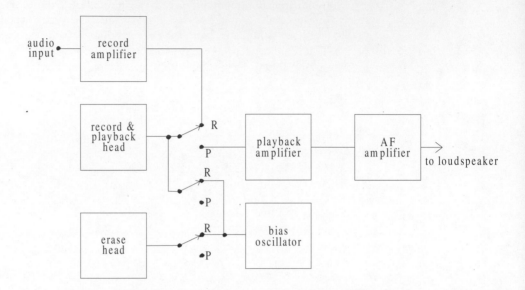

Figure 16.22 The block diagram for a tape-recorder system. The switching is often elaborate, and great care has to be taken to avoid unwanted feedback between different parts of the switch. In the diagram, R and P mean *record* and *playback* switch positions, and all switches are shown in the *record* position

In most tape recorder designs, the record and playback amplifiers make use of the same amplifying stages and the same head, with only the frequency correction networks being interchanged by switching. Several lower-priced designs also make the loudspeaker power output stage serve as the bias oscillator during recording. Figure 16.22 shows a block diagram for a typical tape (which includes cassette tape) recording/replay system.

Question 16.5

In use, the gap in a tapehead will increase. What effect does this have on the performance?

General fault finding

Fault finding for electronic/electrical equipment is a skill that is neither an art nor a science, but an engineering discipline in its own right. Effective fault finding requires:

> A good general knowledge of electricity and electronics.
>
> Specialized knowledge of the faulty equipment.
>
> Suitable test equipment.
>
> Experience in using such test equipment.
>
> The ability to formulate a procedure for isolating a fault.
>
> The availability of service sheets and other guides.

A good general knowledge of electricity/electronics is an essential because not all equipment is well-documented, and in some cases only a circuit diagram (or even nothing at all) may be available as a guide. Failing a concise description of how the equipment works, you may have to work out for yourself the progress of a signal through the equipment. In addition, a wide general knowledge is needed if you are to make reasonable assumptions about how to substitute components. You are not likely to know why something doesn't work if you don't know what *does* make it work.

Specialized knowledge can greatly reduce the time spent in servicing, and if your servicing is confined to a few models of equipment you are likely to know common or recurring faults by their symptoms. All too often, however, service engineers are likely to have to struggle with unfamiliar equipment for a large portion of their time.

Suitable test equipment is essential. The days when a service engineer could function effectively with little more than a multimeter and a screwdriver are long gone, and though the multimeter is still an important tool (as also is the screwdriver) the service engineer needs at least one good general-purpose oscilloscope, along with signal generators, pulse generators, and more specialized equipment appropriate for the type of equipment he/she is working on.

Experience in using test equipment is also an essential. All test instruments have limitations, and you must know what these are and how you can avoid being hung up by these limitations. You must know which tests are appropriate for the faulty equipment, and what the result of such tests would be on equipment that was not faulty.

The ability to formulate a procedure for isolating a fault means that you need to know what to test. All electrical and electronic equipment consists of sections, and much modern electronics equipment uses a single IC per section. You should be able to pin down a fault to one section in a logical way, so that you do not waste time in performing tests on parts of the circuit which could not possibly cause the fault. The classical method of isolating a fault has, in the past, been to check signal inputs and outputs for each stage, but this is no longer the only method that needs to be used, and in some cases, the use of feedback loops, limiters, and other interacting circuits makes it much more difficult to find where a fault lies. Once again, experience is a valuable guide. For white goods, the main test is the presence or absence of mains voltage at critical points.

Basic fault finding in both analogue and digital systems follows principles that are similar. A source is required to inject suitable signals into the input and the signal processing is then monitored as it passes through the system on a stage by stage basis. For analogue systems a suitable input source is a signal generator, while an oscilloscope can be used as a monitor. For digital systems this end to end technique can be carried out using a logic pulser to provide the inputs while the processing can be monitored with a logic probe. A technique that is often used with speed advantage is known as the half split method. Here the system is divided into two sections and the end to end technique used to find the faulty half. This process is repeated continually until the faulty stage is identified.

The availability of service sheets and other guides is also important. Much commercial equipment consists of components which carry only factory codes, and whose actions you can only guess at in the absence of detailed information. In addition, good service sheets will often carry a list of known recurring faults, and will also give valuable hints on fault-finding methods.

Finding a fault is not, unfortunately, a certain step towards repair. Some equipment carries ICs which are no longer in production and for which no replacement is available. Many firms, particularly manufacturers of domestic electronic equipment, will provide spares and help for only a limited period, and some firms seem to deny all responsibility for what their equipment does after a few years. Given the comparatively long life of most electronic equipment, it would be unreasonable to expect spares to be available indefinitely, but it is not easy to tell a customer that the TV receiver he/she bought only six years earlier cannot now be serviced because it contains parts for which there is no current equivalent. Manufacturers might like to remember that customers tend to have long memories about such things – it certainly affects my judgement when I want to buy anything new.

Activity 16.2

Using an AM/FM battery operated portable receiver and an AM/FM signal generator, provide a suitable level of input signal at intermediate frequency to the mixer stage output. Using a double-beam oscilloscope, monitor the signals present at the input and output of the detector stage. Sketch these waveforms and compare their time and amplitude relationships. Switch both receiver and signal generator to FM and repeat the exercise. Again compare the waveforms as before, but also compare the AM and FM waveforms.

Home audio systems

Figure 16.23 shows the major elements of a typical audio amplifier system designed for high quality performance. It is capable of handling a wide range of different input devices ranging from a high grade vinyl disc player to the audio input from a TV receiver to form a complete home cinema installation. Because the many different input transducers have different amplitude and frequency characteristics, the initial input stage will have individual sockets with frequency compensation circuits (equalizer stages) to match the parameters to those of the amplifier input pre-amplifier.

Input signal frequencies will cover between 20 Hz and 20 kHz at signal levels ranging from about 5 millivolts to perhaps 100 millivolts. The power output levels will range from about 100 mwatts for the drive to a pair of headphones for personal listening up to perhaps 100 watts per channel for family entertainment. The demands upon the power supply section will therefore be quite complex. Because of this, the output load current and voltage will be continually monitored by an electronic circuit to provide protection under overload conditions.

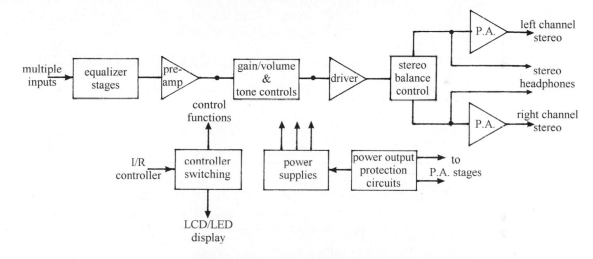

Figure 16.23   Block diagram for a home audio system

Because the amplifier can be used under a wide range of different applications, the input/output devices may be controlled via a switching stage which itself may well be controlled via an infrared remote system. This stage may also include an LCD/LED display panel.

Typical driver input devices for such an audio system may include the following:

Radio tuner unit

This is most likely to provide both AM and FM broadcast signals, but only the latter can be described as producing anything approaching high fidelity (hi-fi) quality. The latest types of digital broadcast tuners (DAB) will produce by far the best quality.

Analogue tape recorder input

In this class, the compact cassette type of device will generally produce the lowest audio quality. The open reel machines which are capable of hi-fi reproduction are now only occasionally found in broadcast studios. From the reliability point of view, both of these devices tend to produce more faults in the drive mechanisms rather than in the electronics sections.

Compact disc (CD) player

These devices are basically digital in function and are therefore capable of producing an output quality that sets the standard for hi-fi against which other systems can be compared. In particular, the lack of background noise from these devices really has to be listened for to be appreciated. The mini-disc (MD) is a recordable version of CD, using a 3-inch disc, and with digital compression used. The quality of MD recording is better than that achieved by cassettes, but not acceptable for the label of hi-fi.

Transcription unit.

These are turntable devices that are intended for the replay of vinyl discs. They are now usually found only in broadcast studios or the homes of enthusiasts. Although quality can be very good, most of these devices tend to suffer from rotation noises, rumble and background hiss. The pick-up transducer, tracking arm and rotational mechanism tend to be somewhat delicate and therefore prone to physical damage. Often moving coil pick-ups are used in these applications. These have a low impedance (50–75 Ω) and generate less than 10 mV of signal at 1 kHz. They therefore need to work into a low noise amplifier that is usually equalized to RIAA (Radio Industry Association of America) standards.

Installation criteria for audio systems

For hi-fi systems it is important that the listening experience should be maximized. This may involve consideration of the room furnishings because hard walls can create unpleasant reflections that exaggerate the reverberation and curtains and similar soft materials can produce unwanted damping effects. Stereo loudspeakers should be carefully positioned to ensure best quality of signal distribution throughout the listening space and this will usually be a compromise. The inter-connecting cables should be as short as possible and routed so as not to create aerial effects that give rise to interference. The signal level settings for each input device should be carefully adjusted to avoid overloading and under-driving.

Answers to questions

16.1	8.8 cm.
16.2	The AF signal will be feeble compared to noise.
16.3	1.495 MHz and 1.505 MHz.
16.4	There is no need to use a high-power linear amplifier stage.
16.5	The effect is to reduce the treble response so that the sound is muffled.

17 Television technology

UK TV system

The PAL I system television as employed within the UK employs amplitude modulation for the vision carrier and frequency modulation for the monophonic sound channel. The sound carrier is positioned 6 MHz above the vision carrier. In addition to these, the carrier for the NICAM stereo version of the same audio programme is located 6.552 MHz above the vision carrier. This stereo information in digital format is superimposed on the subcarrier using a modified form of phase modulation. To avoid adjacent channel interference, this wide band signal is allocated a transmission channel that is 8 MHz wide. The current digital terrestrial television (DTV) service occupies the same segments of the frequency spectrum and these channels are allocated so as not to produce mutual interference.

The UHF part of the frequency spectrum is sub-divided into two bands, Band IV from 470 to 582 MHz and Band V from 614 to 854 MHz, whilst Channel 5 occupies the range from 583 to 599 MHz. Conventionally, these services are covered using Yagi arrays as shown in table 17.1.

Table 17.1   UHF TV aerial groupings.

Channels	Group/Band	Sub-band	Colour
21–37	A	Band IV	Red
35–53	B		Yellow
48–68	C/D		Green
39–68	E	Band V	Brown
21–48	K		Grey
21–68	W		Black

Satellite TV reception which occupies microwave frequencies, is almost entirely via reflector antennas, and operates in the so called C Band, Ku Band and Ka Band. The former band is generally only used in North America, whilst Ku Band carries the bulk of the television services within Europe with a channel spacing of 19.18 MHz. These bands are employed as shown in Table 17.2.

Originally Ku Band was further sub-divided into the FSS band (Fixed Satellite Services) for general telecommunications business, and the DBS band (Direct Broadcasting by Satellite). However, the two segments have merged and together both are often described as providing the DTH (Direct to home) service.

Table 17.2   Satellite television channel allocations.

Band	Up-link (GHz)	Down-link (GHz)
C	5.925–6.425	3.7–4.25
Ku	12.75–13.25 & 14–14.5	10.7–11.7 & 11.7–12.75
Ka	27.5–31.0	17.7–21.2

Question 17.1

For the UK PAL television systems, state the frequency separation between:

(a) Channel 24 sound and vision carriers,

(b) Channel 24 and Channel 25 vision carriers,

(c) Channel 24 sound carrier and Channel 25 vision carrier.

Antennas or aerial systems for television reception

Yagi arrays

The basic half-wave dipole has a circular radiation/reception pattern broadside on its orientation. By adding directors in front and reflectors behind, the so called polar diagram becomes almost pear shaped. The new structure that is now referred to as a Yagi array, thus has a preferred direction of transmission and reception, making it highly directive and selective in the signals that it will process.

| 0 dB | +3.5 dB | +5 dB | 75Ω | 300Ω | 1200Ω | more than 300Ω | less than 300Ω |

(a)   (b)

Figure 17.1   Yagi gain and impedance

Figure 17.1(a) shows how these parasitic elements modify the gain of the array. The dipole is usually used as a reference gain against which other aerials can be measured. Thus a three element Yagi will be quoted as having a gain of 5 dBd (5 dB with reference to a dipole). By comparison, an 18 element array can have a gain as high 18 dBd. The addition of these parasitic elements has the effect of lowering the dipole impedance (typically 75Ω), but this can be countered by modifying the dimensions of the dipole as indicated by Figure 17.1(b). These arrays are commonly used for frequencies up to about 1 GHz. At installation, the aerial array is rotated so that it picks up the maximum wanted signal level, consistent with the minimum of noise.

Reflector antennas

These devices are more commonly used for microwave frequencies, such as satellite transmission and reception. The common type of antenna is shown in Figure 17.2(a), and consists of a dish shape referred to as parabolic. The relationship between the dish shape and its diameter is also shown. For such a shape, any wave or ray emanating from point **a** the focal point, will reflect off the dish parallel to the X-axis. The important feature of this particular shape is that the distance from the focal point to the aperture plane through reflection is constant. Thus wave energy emanating from point **a** will pass through the aperture plane totally in phase. Conversely, any wave energy received along a complimentary path will be concentrated at the focal point **a**. It is this feature that is responsible for the very high gain which increases with the dish area and the operating frequency.

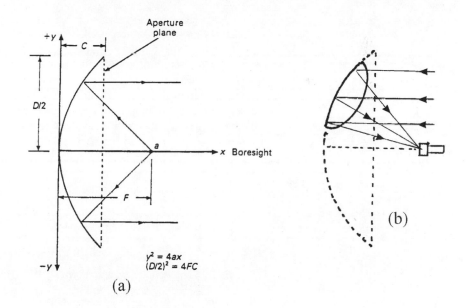

$$y^2 = 4ax$$
$$(D/2)^2 = 4FC$$

Figure 17.2 Reflector antennas

The major problem with these antennas is that the support structure and the radiating or receiving unit located at the focal point actually blocks some of the energy arriving at the dish. For a small dish, this can reduce the overall gain significantly. If a circular section is cut from a larger parabolic reflector, the focal point is no longer in front of the dish as shown in Figure 17.2(b). Since the blocking effect has now been eliminated, the overall gain is higher. The apparent boresight is offset from the true angle by typically 28° and this has to be taken into consideration during installation.

Colour

The sensation that we call *colour* is the effect on the eye of the different frequencies of light rays. The pure white light of the sun is a mixture of all the visible frequencies of light (along with a large range of the invisible frequencies).

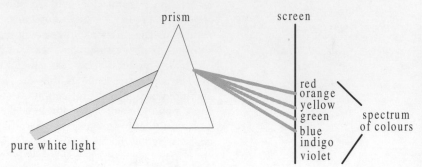

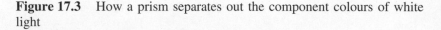

Figure 17.3   How a prism separates out the component colours of white light

These different frequencies can be separated out from white light by the refracting action of a wedge-shaped piece of glass called a *prism*. A prism produces a spectrum of colours, in the following (ascending) order of light frequency: red, orange, yellow, green, blue, violet (indigo also is sometimes identified as existing between blue and violet). Light frequencies lower than red are called *infrared*: those higher than violet are called *ultraviolet*, and both are invisible to our eyes.

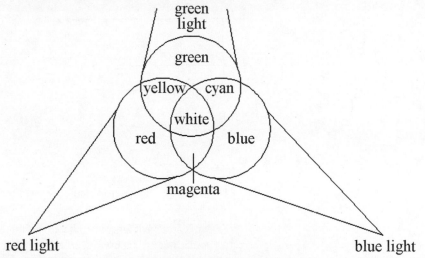

Figure 17.4   How secondary colours are obtained by mixing

A pure white light can be obtained by mixing together in the correct proportions only three colours of the spectrum rather than all of the possible colours. These three colours are called *primary colours*. The primary colours used in colour TV and some colour photographic processes are red, green and blue. The mixture of these three in the correct proportions gives a good quality of white light. By appropriate choice of the standard frequencies of the primary colours, a wide range of secondary colours can be obtained by mixing. Some samples of secondary colours obtained in this way are shown in Figure 17.4.

You can see that yellow can be obtained from an appropriate combination of red and green, cyan from green and blue, magenta from red and blue and white from a combination of all three primary colours: red, green and blue.

Secondary colours can be obtained either by adding primary colours together (in what are called *additive mixers*) or by subtracting primary colours from white light (in *subtractive mixing*). Additive mixing is the process used in colour TV; subtractive mixing is used in most colour photographic systems. A triangle diagram, see left, is often used to show colour addition effects.

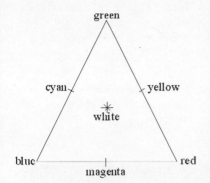

The two important quantities required to describe a colour exactly are its *hue* and the degree of its *saturation*. Hue is specified by the proportion of the primary colours which are present in the colour, and describes the colour itself. Saturation is a measure of the amplitude of that colour. Desaturation of a colour with white light produces pastel shades. Saturated colours are never seen naturally, but they can be used in cartoons and in signals derived from computers.

For colour television the hue and the degree of colour saturation of every part of the picture is defined, as well as its brightness (luminance) and its exact position in the scene being televised. In the colour TV system, the colour information signal is referred to as *chrominance*. The *composite video* waveform from a TV camera contains both the luminance (black and white) signal, and the chrominance (colour) signals.

TV signals

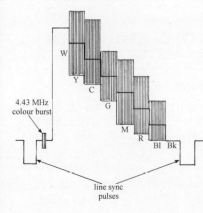

The luminance, luma or Y signal component is formed by adding weighted values of the three camera voltages, Red (R), Green (G) and Blue (B) in the approximate proportions of $Y = 0.3R + 0.6G + 0.1B$. Colour difference signals are then produced to obtain (R–Y) and (B–Y) signals that form the chrominance or chroma component of the signal. These two components are then separately amplitude modulated on to two quadrature (90° phase difference) versions of the same subcarrier frequency and these are added to produce an analogue quadrature amplitude modulated (QAM) signal.

The chroma sub-carrier frequency is very precisely chosen so that the modulated chroma signal can be added to the luma component as shown in the drawing, left. In the PAL system this subcarrier is set at 4.43361875 MHz and this allows the luma and chroma components to be separated at the receiver without mutual interference. Because the green (G) signal component is contained within the Y signal, processing in the receiver decoder recovers the third colour difference signal G–Y. The original R, G, and B signal components are then recovered simply by adding the Y signal to each colour difference. Line synchronizing pulses and a 9 cycle burst of subcarrier are then added to produce the composite video waveform as shown. It is this signal that is used to amplitude modulate the final radiating RF carrier.

Activity 17.1

Use a suitable colour TV receiver fed with a standard colour bar test signal and note the order of the colour bars. (This is usually, white, yellow, cyan, green, magenta, red, blue and black, reading from left to right.) In turn, switch off the red, blue and green guns, both singly and in pairs and record the colour bar patterns obtained. Compare the findings with Figure 17.4.

- It is important to recognize that dangerously high voltages will exist in certain areas of the receiver. Great care should therefore be exercised when carrying out this experiment.

Question 17.2

Deduce the effect on the White, Yellow, Magenta, and Red colour bars if the Red drive signal is missing.

Scanning

Scanning, see Figure 17.5, is a method of obtaining a *video* signal at a fixed repetition frequency from a light-pattern or picture. A TV camera tube or sensor produces at its output a signal which is proportional to the brightness of the light reaching its front surface. A camera tube contains an electron beam which is focused to a spot. The position of this spot on the front surface of the tube is the point at which the brightness is sampled by scanning to produce an output signal. A solid-state sensor is sampled by activating a set of light-sensitive cells in turn. A lens is used to focus an image on the front surface or faceplate of the camera tube or sensor.

The scanning samples all of the surface of the camera tube or sensor. This is done by moving the sample point horizontally (line scan), at very high speed, across the sensitive surface and at the same time deflecting it down the surface (field scan) from top to bottom at a much lower speed.

The signal output from the camera tube or sensor has a varying amplitude that represents the brightness of every single tiny area surface as it is scanned. The signal contains two main frequencies – the line scanning frequency and the field scanning frequency. Sets of pulses, the sync. pulses, are added to this video signal so that the scanning circuits at a receiver will scan at the correct speed and in step with the signal.

odd-numbered line

even-numbered line

1 field = scan of either odd- or even-numbered lines

1 frame = 2 fields, containing all even and odd lines

Figure 17.5   The principle of scanning. TV scanning is interlaced – the odd numbered lines are scanned, followed by the even-numbered lines, so that two scans are needed to cover the screen area

For colour TV signals, three camera tubes or IC light detectors are used for the three colour signals that are needed. Because the three primary colours of light (red, blue and green) add up to white, the signals from the tubes can be changed into a black and white (luminance) signal and two colour signals. The *luminance* signal can be displayed on a monochrome receiver (which ignores the separate colour signals).

Aspect ratio

This term refers to the width to height ratio of the displayed image and almost throughout the lifetime of television, this has been fixed by international agreement at 4:3, largely for the convenience of the CRT manufacturers. Because of the introduction of better transmission systems and alternative means of delivering high quality images direct to the home, this is now in the process of change. The concept of widescreen TV using larger, flat faced tubes with an aspect ratio of 16:9 produces a new and improved viewing experience approaching that of a cinema presentation. For the older receivers left with the 4:3 aspect ratio, the new format is displayed with very noticeable black bands at the top and bottom of the picture. Because this so called *letter box effect* can be disconcerting, a compromise is in operation using an aspect ratio of 14:9 which reduces the height of the banding effect by about 50 per cent. In the true wide screen receiver, a wide screen switching (WSS) flag byte is carried within Packet 30 of the Teletext signal. This automatically signals to the receiver to set the amplitudes of the field and line timebases to suit the programme being broadcast.

Colour CRT

Colour TV reception depends on the use of a colour CRT, or an equivalent display device such as a colour LCD screen. A colour TV tube works on the principle of additive mixing of colours. The three primary colours, red, green and blue, are created at the screen by using three electron guns that are fed with three separate brightness signals. A metal grille, the shadowmask, is fixed close to the screen. This forms a pattern of slits in the metal, and the mask prevents the beam from one gun from striking more than one of the three appropriately coloured phosphor stripes in each group of three stripes on the screen. The beam from one gun lights only the red stripes, the beam of the second gun lights only the blue stripes and the beam of the third gun lights only the green stripes. The sets of stripes are so narrow that when the beam lands on a stripe it causes a small coloured dot to appear, and the picture looks continuous when it is scanned by the fast-moving beam, and when viewed from a reasonable distance from the screen.

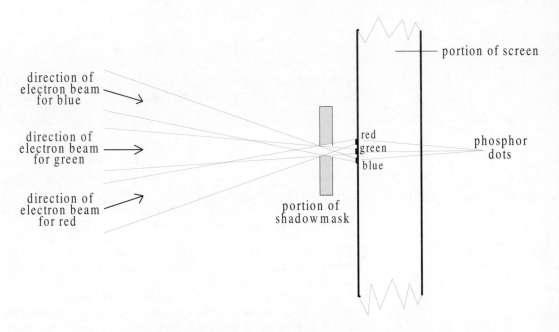

Figure 17.6   How a colour TV receiver tube works

Every detail of a colour picture can therefore be reproduced by the beams from the three guns, using voltages at the cathodes of the gun which are in the correct proportions to give the brightness and colour required as the beams are scanned over the face of the tube.

The beam intensity is varied by altering the voltage to a set of electrodes, the control grids, in the tube. This control is, in practice, a simple potentiometer that alters the d.c. voltage on the grids. The beams are focused by using a focus coil around the neck of the tube. The current through this coil changes the focusing of all three beams. The scanning is carried out using another set of coils, the *yoke*, that carry signals generated by the timebase circuit of the receiver. The changing current through these coils will deflect the beams at a steady rate, and return the beams rapidly at the end of each scan. There are two sets of deflection coils in the yoke, one for horizontal (line) scanning and the other for vertical (field) scanning. Additional coils or magnets are used for *convergence*, meaning that they ensure that all three beams meet at a single area on the screen.

Typical CRT faults include the following:

- Purity faults, when the beams fall on the wrong stripes. A purity fault is obvious when one gun alone is used. On a white screen picture, the single colour will have patches of other colours.
- Convergence faults, which cause colour fringes to appear. This is particularly noticeable on a black and white picture.
- Poor grey scale, because the guns are not equally sensitive.

Many faults can be caused in an old CRT when one gun has low electron emission or fails altogether. Receivers provide for switching each gun off so that the effect can be checked.

Activity 17.2

In the following actions, pay particular attention to safe methods of working and precautions to avoid tube implosion. In addition, it is important to follow the guidelines as presented in the appropriate service manual.

a   Carry out the operation of replacing a cathode ray tube.

b   On a colour TV receiver, carry out the degaussing operation and purity adjustment.

c   Using a pattern generator, carry out the convergence operation, following the manufacturer's service data.

Convergence and purity adjustments may be limited with the more modern receivers. This is because the scan coil assembly, and focus and convergence components will have been matched to the tube during manufacture and then fixed firmly into position.

Effect of electrode voltages on beam control

The decoded video signal is usually applied to the cathodes and the level of this signal affects the picture contrast. In the case of the colour tube, the level of the three signals, Red, Green and Blue need to be accurately matched.

(a)   The overall brightness is dependent upon the level of the d.c. setting of the grid voltage.

(b)   The 1st anode normally operates at a fixed d.c. voltage level.

(c)   The variable focus anode voltage provides a means of obtaining overall image sharpness.

(d)   The final anode voltage is chiefly responsible for the overall image brightness.

Warning. *Whilst most voltages within a television receiver can be measured with a general purpose multi-range meter, the EHT voltage should only be measured using a special high impedance electrostatic volt meter.*

Table 17.3   Comparison of monochrome and colour TV tube voltages.

CRT element	Mono tube	Colour tube
Heater	6.3 or 12 V	6.3 or 12 V
Cathode	+80 V	+150 V
Grid	+20 V	+40 V
1st anode	+300 V	+1500 V
Focus anode	up to 350 V	up to 4 kV
Final anode (EHT)	10 to 15 kV	25 kV

Receiver block diagram

A simplified block diagram of an analogue colour TV receiver is shown in Figure 17.7. The modulated carrier signal, which is at a frequency in the range 400 to 900 MHz, is picked up by the Yagi aerial and linked to the tuner unit by cable. In the tuner, the required frequency band (which will be about 8 MHz wide) is selected, amplified in an *RF amplifier*, and mixed with a sine wave in the usual way. This *mixer* action produces an intermediate-frequency signal occupying a frequency band of 33.5–39.5 MHz. The mixer is tuned using *varactor* diodes so that the alteration of tuning is achieved by altering the d.c. bias on these diodes.

The wide range of frequencies is needed because of the large bandwidth of the video signal, and also because the sound signal is included in the range. For satellite broadcasts, the very weak SHF signals are picked up on a microwave dish aerial, and converted to UHF by a mixer circuit in the dish itself.

The IF signal is then amplified in the IF amplifier. The AFC (automatic frequency control) circuit obtains a d.c. signal from the IF signal, and this d.c. signal is fed back to correct the frequency of the oscillator in the event of frequency drift. The use of AFC enables a station to be tuned in and kept correctly tuned.

AGC is also used to maintain the amplitude of the received signal as constant as possible. The signal from the IF amplifier is then passed to a vision demodulator. The *luminance signal*, which forms the shape and grey tones of the picture, has been amplitude-modulated on to the carrier, and is recovered by a simple amplitude demodulator in the receiver.

The sound signal has been frequency-modulated on to a separate carrier separated by 6 MHz from the vision carrier. Because of the frequency modulation of the sound signal, it cannot be demodulated by the amplitude demodulator; but a mixing action takes place which produces a frequency-modulated 6 MHz signal called the *intercarrier* signal. This signal is amplified by the intercarrier amplifier, and then frequency-demodulated to produce the audio frequency (AF) sound signal which is amplified and applied to the loudspeaker. The volume control is incorporated into the voltage amplifier part of this audio section.

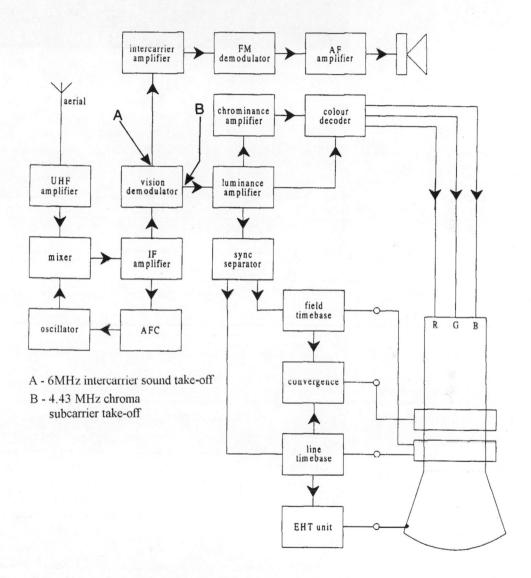

A - 6MHz intercarrier sound take-off
B - 4.43 MHz chroma
   subcarrier take-off

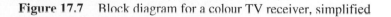

Figure 17.7 Block diagram for a colour TV receiver, simplified

For satellite broadcasts, the very weak microwave signals are picked up by the dish antenna, focused on to the low noise block converter (LNB) for preselection, amplification and conversion to frequencies in the range 950 to 2050 MHz. These signals are then processed through a set top box (STB) or integrated receiver as a first IF.

Activity 17.3

Trace the signal through a working TV which is isolated from the mains supply by a transformer.

Luminance and timebases

To return to the vision demodulator, the vision signal at this point consists of the luminance signal, a waveform whose shape depends on the picture information but which has a repetition rate of 15.625 kHz and a wide bandwidth, together with a modulated subcarrier at about 4.43 MHz. The luminance signal is the signal which carries information about shape and shade, while the modulated subcarrier at 4.43 MHz carries the information about colour. At this stage, both the 4.43 MHz chroma subcarrier and the 6 MHz sound IF signals, can be extracted.

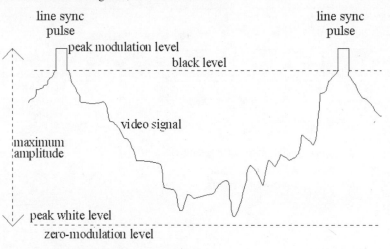

Figure 17.8 The line synchronizing pulses and video signal. After one field of lines a set of sync pulses is sent. This can be separated out and used as a field sync pulse

The luminance amplifier increases the amplitude of these waveforms for the next stage. Part of the luminance amplifier signal is taken to the sync separator in which the recurring synchronizing pulses (see Figure 17.8) are separated from the luminance signal and also from one another. The field sync pulse at a rate of 50 per second is used to synchronize the field timebase. This drives the field scan coils and so deflects the CRT spots vertically from top to bottom of the screen face.

The field timebase circuit drives a low frequency sawtooth waveform (50 Hz) through the low impedance scan coils which behave in a resistive manner. Since the peak-to-peak current is in the order of 3 A at a maximum of about 25 V, the circuit behaves very much like an audio amplifier system which can be incorporated within a single IC as shown in Figure 17.9(a). A protection circuit is provided that will shut down the tube voltages in event of the timebase failure to prevent burning of a line across the tube phosphors. Linearity of scan is assured by using a negative feedback loop that encloses the field scan coils. The circuit, Figure 17.9(b) typically includes two preset controls, one to set the field oscillator frequency and the other to adjust the amplitude or height of the scanning waveform.

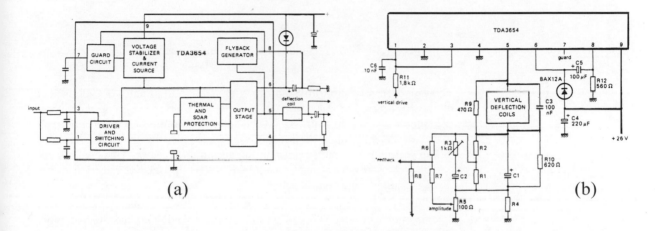

Figure 17.9 (a) internal block diagram of field processor IC, (b) associated circuitry for vertical output

The line sync pulse is used to synchronize the line timebase. It runs at 15.625 kHz, and drives the line scan coils which deflect the spots across the screen to form lines. The action of this timebase produces a pulse of high voltage which is stepped up still further by a transformer, and then rectified to produce the EHT (extra high tension) voltage which is needed to accelerate the electron beam towards the screen of the tube. An EHT voltage of around 24 kV is needed. This is much higher than the EHT of a monochrome receiver, and much of the beam energy is wasted as heat because of the electrons that strike the metal grille instead of the screen.

The line scan circuit is the much more complex of the two timebases (see Figure 17.10). Largely because of the higher frequency, (15.625 kHz) the line scan coil impedance is virtually entirely inductive and these features combine to produce some very high back e.m.f.s. However, this energy can be recovered and used to power several other sections of the television receiver. The repetition frequency is locked to the line sync pulses by using a phase locked loop (PLL) circuit with a long time constant. Thus the circuit will remain in lock even if the sync pulses are lost for a period of time. The line scan power can be quite high, with peak currents of about 6 A at more than 1000 V. The important features of this circuit are the high levels of radiation that can arise which requires effective screening. This in turn, leads to insulation problems and the provision of adequate ventilation.

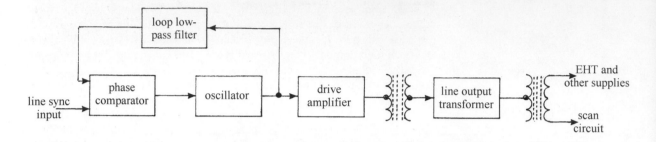

Figure 17.10 Block diagram of line timebase circuit

Preset controls are provided for frequency, amplitude or height and linearity. Frequency control is usually affected through a dust cored inductor associated with the oscillator stage of the PLL and amplitude usually depends upon the level of signal provided by the drive amplifier. The scan linearity can be preset using a saturable reactor. This consists of a coil and permanent magnet assembly where the relative position of the magnet affects the control action.

Because the line output stage runs at a high frequency and with a switching action, this stage is commonly linked with a switched mode power supply that powers the rest of the receiver.

The 4.43 MHz subcarrier signals extracted at an earlier stage are further amplified in the *chrominance* amplifier. They are then demodulated by mixing them with sine waves of exactly the subcarrier frequency and in the correct phase. This demodulation (called synchronous demodulation) produces two signals which are called *colour-difference signals*. If there is no subcarrier, a *colour killer* circuit biases off all the colour signals

The colour-difference signals are then mixed with the luminance signal in part of the colour decoder stage, in which signals representing two of the three basic colours – red, green, blue – are subtracted from the luminance signal to produce the third colour signal. These three separate colour signals are now applied to the appropriate cathodes of the colour tube. In this way, every part of the picture has the correct luminance (brightness) and colour balance.

The highly-simplified account of colour TV reception given above applies to all three of the colour TV systems used in the world – NTSC in the USA, SECAM in France and Russia, and PAL in Britain, Germany and the rest of Western Europe. The differences between the three systems concern only the chrominance amplifier and the colour decoder blocks.

Integrated digital receivers are now available for receiving the new services which provide up to 6 channels on each main UHF carrier frequency. Set top boxes (STB) can be used to convert earlier analogue receivers, but strange interference problems are beginning to show up particularly when a video recorder is added to the system.

Activity 17.4

Using manufacturers' service information and employing safe working practices, carry out the adjustments to the field and line scan time bases.

Installation and servicing

Until recently, the TV receiver installation consisted of little more than a VCR, plus Teletext decoder, now suddenly a whole host of new features are capable of being integrated. These now introduce the problems of digital versatile disk (DVD) player, surround sound system, the home cinema, internet adapters, modem and other links to the personal computer. These all increase the problems for service departments. Whether in the customer's home or the service department, safety to all concerned is of paramount importance.

Within this complex array of equipment, the TV receiver alone effectively contains its own oscilloscope. The display on the tube face under fault conditions can convey a wide range of useful information about the problems, even before the covers have been removed.

In the past when receivers were constructed almost entirely with discrete components, faults often occurred repetitively giving rise to the term stock faults. Many service departments established a good reputation for fast and efficient repair work based on such work. Today, much of the TV receiver is hidden within a few large integrated circuits which now provide a higher degree of reliability. However, stock faults still occur but much less frequently, so that human memory fails to recognize them as quickly as in the past. Today, it is therefore important to support the service personnel with computer based databases.

Two that have proved to be invaluable are from SoftCopy Ltd, Electronic Publishers, Gloucester, GL53 0NU which are available on floppy disks or CD-ROM and produced in conjunction with *Television* journal produced by Reed Business Publications. Alternatively, EURAS International Ltd, Bristol, BS18 2BR, produce a CD-ROM based system. This is packed with service information, circuits diagrams, hints and repair tips that have been gleaned from more than 400 manufacturers, dealers and service centres.

Many robust, portable battery/mains operated items of test gear are now available for use by the service technician. These range from TV pattern generators, digital oscilloscopes with the facilities for the display of both time and frequency based signals and signal strength meters suitable for selecting wanted transmissions and accurately aligning aerial systems. In addition there is the ubiquitous lap top computer programmed with the service database, but which may also have facilities for use with plug-in cards or adapters that convert it into either an electronic voltmeter, oscilloscope or spectrum analyser.

Video cassette recorder

The recording of analogue video signals on magnetic tape follows the same basic principles as analogue audio recording described in Chapter 16. However, the very wide bandwidth of the video signal imposes many restrictions on this simple statement. The general principle of signal processing is shown in Figure 17.11.

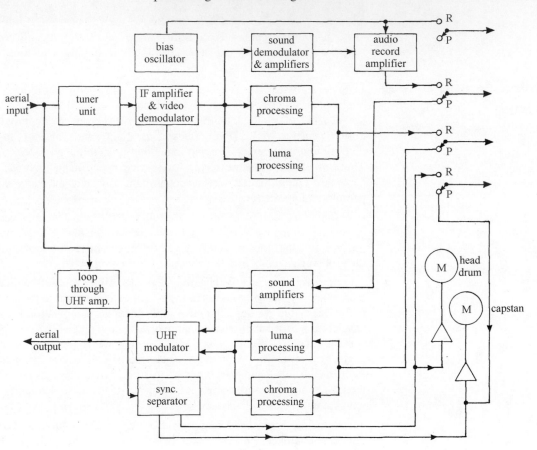

Figure 17.11 Signal processing stages for a video cassette recorder

The high frequencies imply a high head to tape speed and the large amount of information to be stored forces very narrow tracks on the recording medium. These problems are resolved by using rotating record/replay heads which lays down the video signal in a series of very narrow, diagonal stripes across the tape. (About 50μm wide and at an angle of about 5° to the tape edge.) The audio signal can be recorded on a narrow strip along one edge of the tape, whilst field sync pulses are recorded on the opposite edge. The heads and tape travel in the same direction to minimize frictional losses and to extend the mechanical lifetime.

The luminance is recovered from the composite UHF modulated signal and then frequency modulated on to a carrier of about 4.3 MHz, which swings through the range of about 3.8 to 4.8 MHz between sync pulse tip and peak white. Because this FM signal is of constant amplitude, it is used as the bias signal for the chrominance component. This is recovered and down converted to a *colour under frequency* of about 625 kHz when it is added to the luma for recording purposes as shown in Figure 17.11. The replay follows the inverse process, before the luma, chroma and sound are added for remodulation onto a UHF carrier for input to the TV receiver. To ensure synchronism, the tape capstan drive motor is locked to the line sync pulses whilst the head drum rotation is locked to the field sync pulses.

Within the lifetime of this book, the current types of analogue video recorders will be replaced by digital types, using either a hard drive (as used in computers) or a DVD writer drive (also now being used in computers).

Question 17.3

A colour TV is tuned to a very weak station. Why is the picture not in colour?

CCTV

CCTV means closed-circuit TV, and this in turn means that the signals are not broadcast. The camera is linked to one or more video monitors (see later) by cable, so that only these monitors can display the camera picture. In some installations, several cameras can be used, switched one at a time to the monitors.

CCTV cameras are widely used in security systems, allowing several places to be watched in one operation, and the viewing of places that cannot be observed normally. By using night-sight lenses (image intensifiers), pictures can be obtained with very low illumination so that pictures can be obtained when it is impossible to see normally. Infrared sensitive intensifiers can also obtain pictures in very dark conditions.

The least costly types of CCTV systems use monochrome cameras and low-cost TV receivers as monitors. The poor picture quality make these less suitable for recognizing faces, but they are still useful to provide warning of a security risk.

Video monitor

A video monitor consists of a CRT and its driving circuits. The input is a set of video signals, and these are very often the separate R, G and B signals that will be amplified and fed to the cathodes of the (colour) CRT. Video monitors can be used in CCTV, for TV studio use, or for use with computers. For studio and computer use, the picture quality from a monitor must be very much better than can be obtained from a TV receiver, and this quality is expensive. For example, a 21 inch TV receiver could be bought for less than £300, but a monitor of the same size would cost £700 or more, despite the much smaller amount of electronic circuitry in a monitor. The difference is mainly due to the better construction of the CRT

Activity 17.5

Copy the block diagram for a colour TV receiver, and from this create a block diagram for a computer monitor using separate R, G and B signals and separate line and field sync signals.

Monitors for computers are graded by their resolution capabilities. The least costly monitors can attain a resolution of 640 × 480, meaning that you would be also to distinguish separate dots in a pattern using 640 dots wide by 480 dots deep. This is very much better than is possible using a TV receiver, and these monitors are normally used with no interlacing and a field rate of around 80 Hz. The next grade is 800 × 600, and very high-resolution monitors can attain figures of 1280 × 1024 or higher. At the time of writing, monitors using CRT displays are being replaced by LCD types. Note, however, that a change of resolution can have an effect on the display area of the LCD screen. For example, changing a CRT from 1280 × 1024 resolution to 800 × 600 has no effect on the displayed size, but this change made on the LCD display will cause the displayed are to be reduced in proportion.

Answers to questions

17.1   6 MHz, (b) 8 MHz, (c) 2 MHz.

17.2   The following changes occur, White to Cyan, Yellow to Green, Magenta to Blue and Red to Black.

17.3   The colour killer is activated when the subcarrier signal is weak or missing.

Unit 7

Outcomes

1. Demonstrate an understanding of transducers and instrumentation for control systems and apply this knowledge safely in a practical situation.

2. Demonstrate an understanding of control systems and apply this knowledge safely in a practical situation.

18 Control systems and transducers

Terms

A *system* is an assembly of components that is designed to carry out a task. The simplest possible block diagram of any system would show an input block, a process block and an output block. A *control system* is a system that is designed to control some process, so that both input and output are related to the process. For example, a system that controlled water level in a tank would use an input that depended on the water level, and its output would control the flow of water into the tank.

A control system can use an *open loop* or a closed loop. In an open loop system the output has no effect on the process. A water and radiator heating system with no thermostats is an open loop system, because there is nothing that prevents the room from becoming too hot (or remaining too cold). The input (the radiator valves) set the amount of heating (the process), and temperature has no effect on reducing the flow of water in the radiators. When the output (the temperature in this example) can affect the process, the system is a closed loop one. If the radiators are controlled by a thermostat then the valves (or the pump) will switch off when the temperature has reached a set level.

- A closed loop control system uses the same principles as negative feedback in linear amplifiers.

A control system, whether it is of the open or closed loop type, can also be classed as either *on/off* or *linear*. Thinking again of a radiator heating system, if the thermostat is an electrical one that acts on the pump, the thermostat is either on or off. This type of control (also called 'bang-bang control') is simple and often useful, but it less useful if there is a delay in the system.

For example, when the thermostat stops the central heating pump, the radiators are still full of hot water, so that the temperature will continue to rise. The thermostat will not switch the pump on again until the temperature has dropped below the level at which the thermostat switches, by which time the people in the room will feel cold. A graph of temperature plotted against time will show a sawtooth shape that is typical of this type of control.

By contrast, a linear control system regulates the input in a way that depends on the output. Individual radiator valves are of this type, because as the temperature becomes close to the set temperature the valve closes further, and the system will adjust itself so that the temperature remains steady with the radiators just warm enough to maintain the temperature.

Figure 18.1 illustrates the principles of heating systems based on (a) on/off and (b) on linear control.

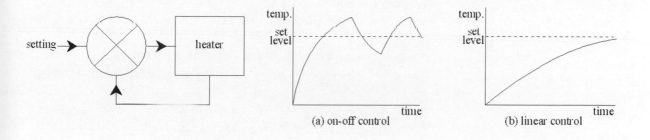

Figure 18.1 Heating systems based on (a) on/off and (b) on linear control

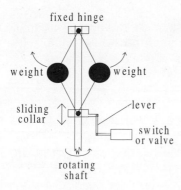

The drawing, left, shows a mechanical speed control system. As the shaft revolves the weights fly out, this movement can be used to operate a switch or a valve. If the switch controls power to an electric motor this is an on/off control, and its speed is unlikely to be steady, it is more likely to rise and fall rhythmically, an action called *hunting*. This is a familiar fault in the regulators for petrol-driven grass cutters and it arises because the system is hunting for a stable condition, something that is almost impossible when on/off control is combined with a closed loop. If the rotating weights operate a valve that will regulate a steam engine, this gives a linear action and maintains speed more smoothly. This is, in fact, the type of regulator that first appeared on steam engines around 170 years ago.

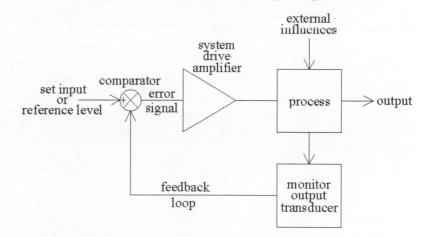

Figure 18.2 A notional closed-loop control system

Figure 18.2 shows a notional closed-loop control system that we can refer to so as to define the terms that follow. There is a set input or reference level, which can be a voltage, that the system uses as its main control. This is compared with an output, using a feedback loop, to generate an error signal. The error signal in turn is used, by way of an amplifier, to control the output, and the action of altering the output will alter the amount of feedback so that the size of the error signal is reduced. Ideally, the error signal would be held at zero, but this is impossible because:

There may be external disturbances that alter the output.

The error signal can reach zero only if the amplifier gain is infinite, which is impossible.

Before a closed-loop control system can be designed, the action must be analysed using four main stages:

measure the output

provide a reference input

generate an error signal

control the process.

The control will normally nowadays be through electrical or electronic circuits, but the output may be anything from packaging control on a production line to environmental controls in an office block. The following terms are important, and should be understood.

Reference or set point input   This is the user input level that represents the desired value of the process variable being controlled. An example is a thermostat setting in a heating control.

Output   This represents the process variable quantity that is to be controlled, such as the room temperature for a heating system.

Comparator   This is the element that generates the error signal by comparing set level with actual level. For a heating system, the thermostat is the comparator.

Error signal   This is the difference between the set point or reference level and the value derived by monitoring the output. It is sometimes referred to as the *system deviation*. For a heating system, this is the difference between the room temperature and the set temperature.

Controller   This is the system element that is driven by the error signal which in turn drives the process in an attempt to minimize the error signal. A room heater is the temperature controller for a heating system.

Proportional control   This is exerted by any system in which the output from the controller is directly proportional to its input error signal.

On-off control   This is a system in which the controller has only two positions, either on or off. Control is thus exerted in a continuous oscillatory manner about some process average. It is sometimes described as bang-bang control. The simple type of room thermostat uses this type of control.

Accuracy   This is the precision with which a system returns to its reference state following some external disturbance.

Offset   Theoretically the error signal should be driven to zero after some disturbance. However, in any practical system, this will not quite be achieved. The difference between the reference value and the final error value is thus described as the offset.

Dead band   This is the range of reference or command values for which the system does not respond. This tends to be greater for electromechanical systems because static friction ('stiction') is higher than rolling friction. To counteract this effect, a low amplitude a.c. signal ('dither') may be added to the d.c. error signal to unstick the mechanism.

Proportional band The proportional band of control is the total range over which the controller is capable of exerting a control over a given system. For example, a temperature controller in a particular system may be capable of operating over a range from 20°C to 100°C, giving a proportional band of control of 80°C. Owing to the proportional gain of this system, the controller may only operate proportionally over a much narrower range, such as ±15°. The system's proportional band would then be stated as 30/80 = 0.375 or 37.5 per cent. Generally, the product of proportional band and proportional gain for a given system tends, within limits, to be constant. A narrow proportional band with its attendant high proportional gain yields a small offset.

Stability The combination of a high-speed response to a disturbance and a small offset can lead to instability. When such a system reacts suddenly to a change, it generates a large error signal that overcorrects. This then produces a further overcorrection in the opposite sense and the system oscillates. During this period, the system is effectively out of control, and unless the oscillation decays, the system will self-destruct. Stability is thus closely related to *damping*.

Damping This is electrically equivalent to shunting a resonant circuit with a resistance or mechanically loading a springy system with a shock-absorber. When a high-gain system experiences a sudden change, there is a tendency for the error signal to oscillate ('hunt') in the manner shown in Figure 18.3. By adding resistance to the control circuit, the response speed is slowed somewhat but the oscillations become controlled. Critical damping is defined as shown, with just one overshoot before the system settles once again to its equilibrium state. Over-damping removes any overshoot but produces a relatively slow response to a change.

Transport lag (transfer lag) For sudden disturbances the controller makes a partial immediate response but the system takes a period of time (lag) to settle down to a new equilibrium. Electrically this is equivalent to the low-pass filtering of a square-wave pulse.

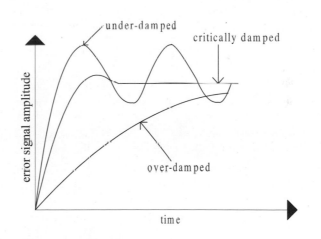

Figure 18.3 The effects of damping on system response. Note that under-damping is associated with hunting

Defining a 'good' system

High proportional gain provides a narrow proportional band to give close control over system variability. Too high a gain can lead to an unstable system. Low proportional gain provides a slow response to any change and a wide range of command levels, but increases the dead band and the offset. A 'good' system is therefore one that:

- is stable and responds quickly to any external disturbance;
- reduces the error signal practically to zero (low offset) to give close control of the process variable.

Question 18.1

When Rudolph Diesel developed his oil engine, he found that simply reducing fuel supply would cause an increase in speed before the engine shut down. How is a diesel engine throttled?

Input transducers and sensors

The input to a system may be of the on/off type, such as a switch. This is often used in electrical systems, such as for positioning a lift, setting the limit of movement for a machine, or cutting off electrical supplies when a safety-cover is removed. An alternative is a linear input, as from a potentiometer. Both switches and potentiometers need protection if the surroundings are *hostile*, meaning corrosive or flammable, · and the applicability of different types of input devices to these surroundings is indicated below.

The microswitch is a small switch that needs only a small amount of movement to switch between on and off. Microswitches are manufactured in vast numbers for use in control systems, and are very reliable mechanisms. The simplest type uses a plunger which is really a thick pin, and a movement of 0.5 to 1 mm of this plunger is enough to switch between on and off. This is used mainly when the system is a mechanical one that will strike the pin straight on. For example, the movement of a sliding door can be used to hit a microswitch pin that signals that the door is fully closed.

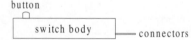

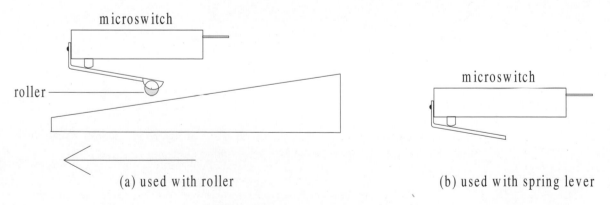

(a) used with roller   (b) used with spring lever

Figure 18.4   Other ways of operating the microswitch

Another way of operating the microswitch is by way of a roller. This is useful when a rubbing contact is made with the switch, and the drawing in Figure 18.4(a) shows this used to detect a level of height along a wedge. Using a roller avoids problems of friction that would make the plunger type of action unsuitable. The other option for microswitches is lever action, indicated in Figure 18.4(b). The level provides some spring action so that excessive movement does not damage the microswitch, and it also allows the movement at the end of the lever to be more than can be used at the switch plunger.

Microswitches cannot be completely sealed, so that they cannot be used immersed in liquids, or in corrosive gases, unless the switch is encapsulated. When plunger operation is used it is often possible to operate the plunger through a flexible covering.

For some systems, a security key switch is appropriate. This is an electrical switch that can be operated only by turning a key, and it prevents the setting from being changed by anyone unauthorized. Security key switches can be used, for example, to set alarm systems, to disable safety systems for servicing, or to allow machinery to be started and used. Because of the need to insert a key, these switches cannot be sealed and cannot be used in a hostile environment.

The oldest type of switch input used in electronic circuits is the straightforward lever switch., whose external appearance is illustrated here, is used for many purposes, particularly for supply units and mains switches. The internal construction uses a toggle arm and a sprint, so that there are two stable positions for the switch contacts, open or closed, with no intermediate position.

Push-button switches can be of two types. The more usual type is push for on and push for off, and this needs some type of indicator to show whether the switch is on or off. Another version uses two switches, one for each action, and again some type of indicator is needed to show the current state. Push-button switches can be operated inside a flexible sealed casing, so that they can be used in liquids and in dangerous gases.

Whichever type of action is used, push-button switching needs a *latching* action that will operate the switch contacts by making use of the mechanical action. This latching action ensures that both switching on and switching off can be performed by using the same circuit. Latching can be mechanical, using a toothed wheel, it can be electrical, using relays (see later), or electronic using a bistable circuit. A push-button switch for use with an electronic bistable is of much simpler construction.

Switches need not use hand action, and foot switches can be used, as they have been for many years to operate sewing-machines. The *tilt switch* will operate when the angle of its body is altered, and one very common design uses mercury in an arc-shaped tube. The mercury provides the switching action, with one contact permanently in the mercury, and the other touching the mercury when the tube is tilted. This can be used to close a circuit when a critical angle is reached, such as the angle of roll of a locomotive or a ship, or it can be used to detect attempts to winch a locked car up a ramp. The tilt switch can be used in a hostile environment, subject

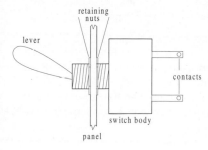

lever
retaining nuts
contacts
switch body
panel

contacts
mercury
close-up view of contacts, end-on

to the temperature limits set by the freezing and boiling temperatures of mercury.

Float switches use the ancient mechanism that is also used to control the level of water in flush toilets. The float and its arm can operate a microswitch to provide on-off control based on liquid level. The float can be of a material that will withstand a corrosive liquid, but the switching part of the mechanism must be protected if the atmosphere is corrosive or flammable.

The switch action need not depend on mechanical movement. A *reed switch* is illustrated here. This uses two thin metal wafers that are the switch contacts, encased in an evacuated glass tube. The wafers can be clamped together, making the electrical circuit, by the action of a magnet, either a permanent magnet or a solenoid, so that the magnet need only be close to the reed switch to operate it. Since all parts of a reed switch like this can be enclosed in plastic or glass it is a useful method of controlling switching in corrosive atmospheres. The reed relay uses more than one set of reed contacts operated by a single solenoid, so that several circuits can be switched by energizing the coil. Reed switches are particularly well suited to hostile environments because the switch contacts and the magnet can be encapsulated.

Switch contacts are subject to the problems of sparking, particularly when an inductive circuit is broken. The rate of breaking the circuit is important, as is the material used for the switch contacts. Switches are usually rated for higher a.c. currents than can be used in d.c. circuits, and a switch that has failed must be replaced by one of the same specification in terms of contact materials and current ratings.

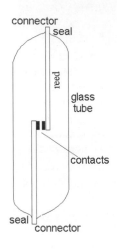

Question 18.2

Why is the a.c. current rating for a switch higher than the d.c. rating?

Temperature

Temperature can be sensed using a *thermostat*. A typical form of thermostat uses a bimetal strip that will bend as temperature changes. If this strip carries a contact it can be used directly to open or close a circuit. The bimetal strip can be coiled to increase the sensitivity. This type of thermostat is very commonly used in air-temperature thermostats for heating systems. This type of thermostat is unsuited to hostile environments.

More precise temperature sensing can be carried out using a thermocouple, see Chapter 13. The thermocouple has a voltage output that depends on the temperature difference between its junctions, and because the voltage is small, a few millivolts, amplification is needed. Thermocouples are more likely to be used when the control system uses electronic circuits. Thermistors (see Chapter 19) have a resistance that changes considerably as the temperature changes, and they are extensively used in electronic temperature sensing equipment. Both types may need protection from corrosive gases or liquids.

Potentiometers provide an output that depends on rotation or position, and are extensively used as input transducers. Sometimes an input needs to be used to set a value, typically a precise voltage. One common method uses a multi-turn potentiometer and dial. The potentiometer shaft can be turned through more than one turn, typically 10 turns, and the dial will register on a scale (such as 0 to 1000). Both analogue and digital dials can be used. This allows for very precise setting, and is used in measuring equipment.

A few types of potentiometers allow continuous rotation, so that the potentiometer output can be used to indicate the angle of a shaft. The types of potentiometer that make use of mechanical contact are not intended for continuous rotation, but the non-contact type can be used on continuously rotating shafts. A potentiometer action can also be linear, using a straight resistor body with a slider that moves over it in a line. This can be used to indicate the amount of movement. In general, potentiometers must not be used in hostile environments.

The strain gauge provides an output that depends on how the gauge, wire or semiconductor, is distorted. This is used to measure the effect of stress (such as pressure on a structure), and is widely used to monitor strain in buildings, turbine blades, bridges and other structures that must not be allowed to fail under stress. Load cells are used to provide a signal that is proportional to pressure, and are used to monitor forces applied to materials.

Output transducers

Output transducers convert signals into other forms of energy, and in control systems we are mainly concerned with transducers that will convert electrical signals into such forms as mechanical movement (linear and rotary), light, or heat. In this section, we are concerned mainly with mechanical actions and with relays that use mechanical action to operate switches. A relay allows a low voltage and low current circuit to switch much higher levels of voltage and current.

Solenoids

Solenoids can be linear or rotary. A linear solenoid is basically a coil of wire wound on a hollow former, with a soft-iron core that can move freely inside the former. The position of the core may be set using a spring. When current flows in the coil, the core is attracted into the former, and its position is determined by the amount of current. This provides a few centimetres of movement, with forces that depend on the amount of power dissipated in the solenoid.

Solenoids can be obtained for mechanical action only, or ready-coupled to devices like valves for liquid or gas control. The specifications for the general-purpose type of solenoids will include both electrical and mechanical ratings, and the solenoids can be classed according to the type of mechanical action as push–pull, lever or rotation.

Another factor is *duty cycle*. Some solenoids may be required to operate continuously for long periods, so that their duty cycle is 100 per cent.

Others may operate only at intervals and for a short *on* time, so that a solenoid that was activated for 30 seconds in each 5 minutes would have a 10 per cent duty cycle. The ratings of a solenoid include allowances for different duty cycles, so that higher voltages can be applied, if required, to solenoids whose action is subject to a low duty cycle.

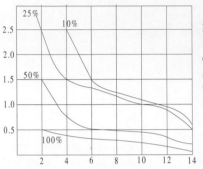

Push–pull solenoids are available as d.c. (12 V or 24 V) or a.c. (110 V or 240 V) operated devices, with a specified coil dissipation such as 10 W or 11 VA for a 100 per cent duty cycle. The amount of force that can be exerted then depends on the *stroke* (the amount of mechanical movement) and the duty cycle, assuming that the maximum operating voltage will be used for each value of duty cycle. These relationships are not simple, and are best expressed in graphical form. In the illustration, left, force is plotted vertically and stroke is plotted horizontally.

The mechanical equipment that is operated by the solenoid should be arranged so that the combination of force and stroke that the solenoid can deliver will be suitable. The fact that the maximum of force can be delivered at the start of an action (when the stroke is negligibly small) means that the solenoid is well-equipped to overcome initial friction. One point to note is that if a.c. operation is used, the mechanism must not be capable of *stalling*; the armature must always be able to travel fully into the solenoid, otherwise an excessive current will be drawn.

Lever solenoids are constructed like relays (see later). These types of solenoids are intended for a range of lower forces and strokes. Like a relay, lever solenoids are equipped with a return spring, though this can be removed if there is a spring action in the mechanism that the solenoid is operating. The *rotary* type of solenoid produces torque (turning effort) rather than force, but has the same shape of graph of torque plotted against rotation for each value of duty cycle. Spring return is normal, but the spring can be removed if there is any other way of rotating the shaft back when the coil is de-energized. The maximum stroke is of the order of 45°.

Motors

D.c. permanent magnet motors are used to a very considerable extent where larger amounts of movement are needed. Another advantage of using permanent-magnet motors is easy reversal of direction of rotation. Miniature general-purpose d.c. motors, using permanent magnet fields, can be controlled by the output from a comparatively low-output semiconductor amplifier. Typically they have voltage requirements of 6–12 V and currents of 200–400 mA. Small electric motors of this type can be used in models, rotating signs, display systems and to some extent in robotics.

Another electromechanical device in the motor family is the stepping (or stepper) motor. The shaft of a stepper motor will rotate by a fixed angle for each electrical input (or set of inputs), and combinations of stepping motor and lead-screw drive can be used to provide linear steps in place of steps of rotation. The principle of stepper motors is multi-phase drive, and a typical stepper would normally use a four-phase set of windings. The stepping action is achieved by pulsing the four windings in sequence, using pulses derived from an IC driver. For example, if the four windings are L1, L2, L3 and L4, then a typical sequence would be:

L1	L2	L3	L4
ON	OFF	ON	OFF
OFF	ON	ON	OFF
OFF	ON	OFF	ON
ON	OFF	OFF	ON
ON	OFF	ON	OFF

Note that two windings change in each step. The cycle as shown here repeats after four steps, and each step would typically take the shaft through 1.8°, so that 50 repetitions of this pattern would turn the shaft through a full revolution. Stepping motors are usually operated at low voltages, 3 V or 5 V d.c., and with fairly low currents in the range 100 mA to 2.0 A. The driver circuits will need to use heatsinks, however, and very great care must be taken to ensure that none of the leads to the motor becomes disconnected while the other leads are carrying pulses, as this will always cause serious damage. Though the *average* currents are fairly low, the peak currents in each pulse can be large. If greater amounts of torque (turning effort) are needed along with smaller steps, gearboxes can be attached, but precision will be lost if this is done because of the inevitable backlash in the gearbox.

Activity 18.1

For a d.c. motor with a wound armature and field, plot the speed/current characteristics for current flow (a) in the armature and (b) in the field.

Relays

Electromagnetic relays make use of the magnetic field in the core of an inductor to alter the setting of a mechanical switch. The essential parts of a relay are a coil with a soft magnetic core, a moving *armature* and a set of contacts which are actuated by the movement of the armature, and which are insulated from the coil and from the metal frame of the relay. The coil circuit is known as the *primary circuit* of the relay, and the contact circuit as the *secondary*. The principles of the traditional form of relay have already been covered in Unit 4, Vhapter 11. Relays can, with some re-design of the magnetic circuit, be driven by a.c., making use of the principle that an armature will move so as to close a magnetic circuit when the energizing coil is driven by a.c.

Many of the applications for the traditional type of relay have been filled more recently by the use of reed-relays and by the use of semiconductor devices such as thyristors.

The power that is dissipated in the operating coil will determine the mechanical force that is available. This is sometimes specified as *pull-in power*, the minimum power needed to make the contacts operate, and the *holding power* is also sometimes specified, the minimum power that is needed to keep the contacts switched over after they have been pulled in. There is usually a significant difference between the two values, so that if a relay is intended to remain activated for a long period after initial activation, then the dissipated power can be considerably reduced by lowering the voltage or current so that only the hold-in power is dissipated. The range of power levels is sometimes quoted in terms of operating voltage for a particular coil, so that a relay nominally intended for 12 V operation might be assigned a voltage range of 9–13 V. Where relays are intended for a.c. operation, nominal volt-amp ratings are shown as well as the usual voltage and current ratings for the coil.

When replacing a relay, the ratings must be identical. Never substitute a solid-state relay for a mechanical one, because wiring regulations prohibit this in many circumstances.

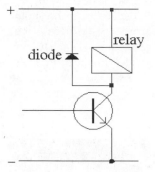

The coil inductance of a relay is seldom stated, but it is important if the relay is driven from a semiconductor device. When the coil current is switched off a voltage pulse is generated, and this can damage semiconductors. A relay coil should never be driven from a semiconductor circuit unless there is diode protection, using the circuit illustrated. Where the relay coil is controlled by a mechanical switch, diode suppression is not strictly necessary, and a series resistor–capacitor suppression circuit can be used. Manufacturers of relays will advise on the suppression circuits that are most appropriate for any particular relay.

Contact configurations are specified in terms of the number of poles and whether these are normally closed (nc), normally open (no) or change-over (co). As for switches, the contact material and current ratings are important, and a replacement relay must use the same type of contacts as the relay that has failed.

Conventional mechanical relays exist in a huge variety of forms, from subminiature types that are enclosed in metal cans of the same size of the TO-5 transistor casing, to heavy-duty types that require to be mounted on a steel chassis.

Latched relays and circuits

A latching relay is one that can be switched on by a pulse of current. It will then remain on until it is switched off by the application of a current pulse of the opposite polarity. Such devices often find applications in motor-switching circuits.

Mechanical latching devices rely on a spring-loaded latch to lock the armature in place after the magnetic action has produced closure. A push-button, or a second coil and armature is then needed to release the latch so that the relay can return to its de-energized position.

Remanence relays either have a small permanent magnetic sleeve mounted on the core, or the core is made of a special ferrous material which retains some magnetism from a current pulse. Once operated, the

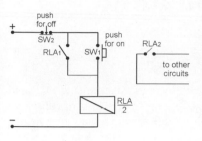

relay remains energized in a power-free mode. Release is obtained by the application of a pulse of the reverse polarity. Operating pulse durations vary between about 10 ms and 20 ms. For a.c. operation, the remanence relay is often driven via series rectifier diodes.

Electrical latching circuits similar to that shown here are more commonly used for these applications. In the circuit illustrated here, when the switch SW_1, is closed, the relay becomes energized; contacts RLA_1 and RLA_2 close so that, when SW_1 is released, the coil remains energized. The circuit will remain latched until released by the operation of SW_2. The use of a relay is indicated on the diagram by two symbols, one representing the coil, the other the contacts. The two symbols often appear in widely separated parts of the circuit diagram, and the drawing convention is therefore known as the detached-contact system, of which this diagram represents a simple example.

Answers to questions

18.1 Diesel engines use a closed-loop control system that uses engine speed to regulate both air intake and fuel injection.

18.2 The contacts may break as the current is minimum and any sparking will quench within a half-cycle of voltage.

19 Instrumentation and control systems

Instrumentation

Instrumentation for industrial control includes many instruments that are familiar from use in other branches of electronics, and also some that are not widely used other than for industrial control. Instrumentation systems are designed for measurement rather than for control, and precise calibration is important. We can look at some examples of typical instruments, starting with a panel-mounting batch·counter.

Batch counters are used to count items passing on an assembly line, and the methods used may be mechanical or electronic. Mechanical counters are suited to slower counting of fairly large objects, and we need not consider them further.

The electronic type of counter will use a count input, fed with pulses for transducers that are operated by the count item. These would be devices such as microswitches operated by objects striking them, or photocells operated by an object breaking a light beam. The input pulse will have to be of a specified peak amplitude and rise time, so that some processing of the signal is likely to be needed. For example, a pulse from a switch will need debouncing, and a pulse from a photocell will need amplification and the use of a monostable to generate a clean pulse. The number of digits used will depend on the likely count (hundreds, thousands, tens of thousands?).

A batch counter of this type will provide the following facilities:

Up/down counting: the direction of counting can be set by a switch to provide for counting up from zero or down to zero from a pre-set amount. The direction of determined by the connections between counter units where flip-flips are used, or by using an alternate piece of program for a microprocessor operated counter

Preset: This allows a count number to be put into the registers and displayed, so that a count down from this number can be carried out.

Reset: This allows the counter units to be set to zero in preparation for a count up of the batch items.

Leading zero suppression: The display will not show zeros ahead of significant figures, so that, for example, a six figure display will show item 1 as 1 rather than as 000001.

Backup: A miniature cell or battery is used to keep the counter operating when the mains supply fails. The backup action usually operates the counter only, not the display, and the cell or battery is maintained charged when the mains supply is on.

Hour meters

Hour meters are used to determine the working time of a system, and are extensively used to determine when servicing or recalibration of equipment is required, or to assess rental cost. The simplest hour meters are electromechanical, based on synchronous a.c. motors as used in electric clocks. The motor is coupled to a mechanical counter so as to provide a digital readout that can be mechanically reset when required. A few older units use a pointer display rather than a digital readout.

Electronic hour meters are based on quartz crystal oscillators and counter units, and can often be switched between a count up or a count down. The display is always digital.

Controllers for systems

A *sequencer* is a device that can be operated to provide a sequence of actions rather than just a single action. For example, the programmer of a washing machine will carry out the actions of operating water taps, pump, washing action motor and spin motor in a sequence that is determined by the program that is being followed. Sequencers can be electromechanical or electronic, and one advantage of using electromechanical sequencers for simple systems is that a failure of the power supply does not reset the sequence.

Electromechanical sequencers, such as the old-type washing-machine programmer, use an electric motor as the driving device to operate cams that open and close switch contacts. They can also allow for other inputs, and for some actions, the motor circuits will be switched off until another input is present. For example, the motor of a washing-machine controller is switched off in the water-heating cycle until the water temperature thermostat contacts close.

Electronic sequencers have replaced the electromechanical type to some extent, depending on the applications. These use electronic timing and switching, though in some cases a relay is used to switch mains a.c. power on and off. Timing for short times can use capacitor charging, like the type 555 timer, but for longer times digital counters are more reliable and precise.

Counters and timers for industrial control are based on digital counting, using a set of bistables. The usual method is to set up a count or time number on the flip-flops and count down to zero, with a switch pulse sent out on the zero count.

More modern equipment is microprocessor controlled, so that programming can be more flexible. This allows also for much longer time periods and much larger counts. The microprocessor is driven by a clock pulse circuit that typically uses a 1 MHz crystal oscillator, so that the precision of the system is determined by the crystal.

- Note that when servicing has been carried out on any counter or timer equipment, the system should be re-calibrated to ensure accuracy.

Programmable logic controllers

Programmable logic controllers (PLCs) are boards or sets of boards containing a CPU, ROM, RAM and EPROM, timers and counters, with provision for input and output of several voltage levels, and a power supply. The CPU for a PLC is likely to be a microcontroller, so that there is often some confusion between PLCs and microcontrollers. In some documents, the terms are used interchangeably. The PLC may be used alone, or as part of a control network under the control of a main computer system.

The extensive use of PLCs dates from about 1969, and before that date the actions that they provide were carried out using relays and hard-wired programmers that used motors, cams, and microswitches. The advantages of using PLCs are:

1. They are more rugged, that is, able to function in the conditions of vibration, noise, adverse temperature and humidity that are likely to be encountered in machine control applications.

2. They are programmable, so that their actions can be changed, enhanced or extended without the need for rewiring or mechanical adjustment.

3. They contain interfacing to allow for easy connection to typical inputs and outputs.

4. They can take up considerably less space than their mechanical counterparts, and do not wear out.

5. Their use considerably reduces the need for elaborate wiring, which also reduces the incidence of wiring errors.

6. The manufacturers of PLCs can provide programming units that allow for programming in familiar high-level computing languages such as C or BASIC or in terms of relay design 'languages' such as ladder diagrams.

Nowadays we would consider a PLC control system ideal for any processes that involved repetitive operations or actions needing timing or which are triggered by external signals. A PLC also scores over any mechanical system when high operating speed is required, and its use is virtually essential if data has to be acquired and/or processed.

A PLC can be represented by a simple block, with one or more inputs and outputs, along with a program input. Inputs will be provided by various switches or sensors, and the outputs will actuate relays, valves, motors etc. The program will consist of a sequence of instructions in which inputs will be sampled, logic actions carried out, and outputs changed. Typically, several parts of the program will consist of loops, meaning that the actions of sampling inputs and carrying out logic actions will be repeated many times until some condition is satisfied. For example, a temperature sensor could be sampled repeatedly until the reading corresponded to some fixed temperature (such as the correct water temperature for washing lightly-soiled cotton).

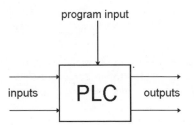

Question 19.1

Why are electro-mechanical sequencing controllers still regarded as more reliable for some applications?

Practical systems
Temperature control

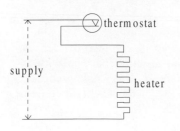

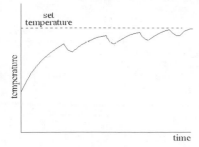

Environmental temperature control can be achieved by using a closed loop system with a thermostat as the control. As the diagram shows the thermostat is wired in series with the heater element. With the thermostat contacts closed, the current flows through the thermostat and through the heater causing the temperature of the environment to rise. When the temperature reaches the set level the thermostat contacts will open and cut off the current. This is a *bang-bang* system.

In practice, it's not quite so simple. A thermostat is a device that exhibits hysteresis so that the temperature for opening the contacts is not the same as the temperature for closing the contacts. This means that for a rising temperature the thermostat contacts will remain closed until the temperature has risen some amount higher than the setting. Similarly when the thermostat contacts are open and the temperature is decreasing the temperature will fall some amount below the set temperature before the contacts close again. This is another example of hysteresis, but in a mechanical form this time.

The practical result of this is that a graph of temperature plotted against time is of the zigzag shape that was illustrated in Figure 18.1. The usual method of adapting a thermostat for better performance is to incorporate a small heating element inside the thermostat so that the heating element passes current while the contacts are closed.

While the thermostat is passing current the action of this heater, called an *accelerator*, is to make the rate of change of temperature higher nearer the thermostat element. This accelerates the action of the thermostat so that the contacts open before the set temperature is reached. The contacts open, and remain open, so that the current no longer flows through the accelerator. This causes the temperature to fall slightly allowing the contacts to close again. The net effect is that the amount of zigzag is reduced as illustrated. left. A less desirable effect is that the thermostat does not reach its set temperature unless the rate of loss of heat is very small. Closer control of temperature can be obtained by using electronic controls along with thermistor detectors.

To control the temperature of a liquid in a process the conventional bimetal thermostat is unsuitable. Instead, a smaller form of detector is necessary, and this can be provided by using a thermocouple or by a suitably encased thermistor. In either case some electronics circuits are needed to amplify the signals and to operate a relay that will switch the current to a heater for the liquid.

Temperature control of a liquid in a process requires a form of temperature detector that will be immune to flow of liquid and to any corrosive effects of the liquid. Various forms of detectors are available in encapsulated terms either in plastic or in stainless steel that allow temperature to be sensed without the danger of corrosion from the liquid.

Liquid level

Liquid level can be controlled using an electrically-operated valve for liquid flow and a float or other liquid sensor. When the float operates its microswitch, the valve is shut off, and when liquid is taken from the tank, the float level will fall and the microswitch contacts will make, operating the valve so that more liquid enters the tank.

Digital weighing

A digital weighing machine provides an output of digital codes proportional to weight. This requires a transducer whose electrical output is proportional to weight along with an analogue to digital converter. For example the output could be obtained from a spring balance with the movement of the balance operating a potentiometer. This would produce an analogue output from the potentiometer and this analogue output could be converted to digital in a D to A converter.

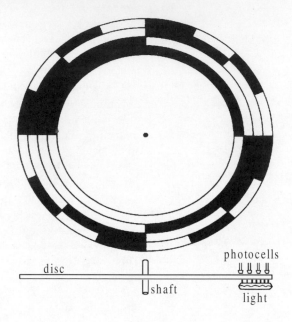

Figure 19.1   An optical digitizer. Each position of the wheel will produce a different pattern of signals from the photodiodes.

Another option is to use the movement of the balance to operate an optical digitizer wheel of the type illustrated in Figure 19.1. As the wheel rotates it produces a digital output directly and since this is proportional to the movement of the balance the output is proportional to weight.

Different methods have to be used for a bulk material weighing system. Bulk weighing might, for example, be via a wagon. Its empty weight is known – this is called the tare weight. The gross weight is then that with load, and the difference is the weight of the material. The actual weighing action is done by using load cells (strain gauges) on the weighing platform (bridge). This method can be extended to material that is passing along a conveyer belt by placing a platform under a part of the belt to sample continuously the weight of load on that portion. Again, the weighing action would be carried out by using load cells.

Filling systems

Automatic filling systems need to be designed to match the material that is used for filling. For example, filling bottles with pills would use a counter to ensure that each bottle contained exactly the same set number of pills. Filling bottles with whisky might be controlled by measuring the level of liquid. Gas filling, as for fire extinguishers, can be done by pressure or by weight measurements.

Preventive maintenance

The electrical or electronic portion of a control system should be constructed from standard units that require little or no maintenance, so that the main problems of maintenance arise from the mechanical parts of the system. In particular, parts that are exposed to the environment need to be maintained on a scheduled basis, and a suitable schedule must be drawn up based on experience.

A considerable amount of preventive maintenance can be avoided by giving some thought at the design stage to the placing of sensors. For example, a temperature sensor placed outside a building can be covered in leaves, paper or other wind-borne rubbish. It can read low because it becomes wet (so that wind evaporating the water cools the sensor), or read high because of direct sunlight.

Most of the maintenance, however, will be directed to moving parts, and in particular bearings, belts, chains, wheels, gears, motors and tacho-generators.

Measurement and fault diagnosis

The servicing of control systems requires instruments. Some of these, such as multimeters and oscilloscopes, will be identical to the instruments that are used for other forms of servicing, but others are designed and used only for process control equipment. For example, the low speed of some systems makes a chart recorder more useful than an oscilloscope for some work, particularly since the output forms a permanent record. For digital signals, an instrumentation tape recorder may be needed, particularly if the action of the system is to be analysed over a long time period.

Function generators are signal generators that are designed to produce sine, square or triangular waves. Each may be varied typically over a frequency range covering from less than 1 Hz to about 20 MHz. The basic frequency may be generated either by a highly stable oscillator circuit, or by using a frequency synthesizer. The basic output signals can be modified. The sine wave can be phase-shifted, and the square wave can have its mark/space ratio varied so that a stream of pulses with a variable duty cycle can be provided. The triangular wave can be varied to give a sawtooth of varying rise and fall times. It is also possible to add a d.c. value to each output to give an *offset*. This is valuable for testing d.c. coupled circuits or other circuits whose frequency response extends down to zero.

Answer to question

19.1   A mechanical sequencer can recover easily after a power failure, but an electronic sequencer needs a backup cell which is itself a component liable to failure. Nicad cells in particular often fail when they have been in use for some time.

Unit 8

Outcomes

1. Demonstrate an understanding of a basic PC system and apply this knowledge safely in a practical situation.

2. Demonstrate an understanding of basic input/output devices and apply this knowledge safely in a practical situation.

3. Demonstrate an understanding of data storage modules and apply this knowledge safely in a practical situation.

4. Demonstrate an understanding of current printers and apply this knowledge safely in a practical situation.

20 The PC computer

PC development

A computer is a logic system that is *programmable*, meaning that the actions it carries out can be altered by feeding in program instructions in the form of codes. The heart of a computer is the *microprocessor*, which is a very complex programmable logic chip.

The type of machine that we now describe as a PC means one that is modelled on the IBM PC (Personal Computer) type of machine that first appeared in 1981. The reason that this type of machine has become dominant is the simple one of continuity – programs that will work on the original IBM PC machine will work on later versions and will still work even on today's machines. By maintaining compatibility, the designers have ensured that when you change computer, keeping to a PC type of machine, you do not need to change software (programs). Since the value of your software is much greater than the value of the hardware (the computer itself), this has ensured that the PC type of machine became dominant in business and other serious applications. Other types of computers are not compatible with the PC or with each other, have less choice of software and more expensive components.

In short, a real PC machine is currently identified by the following points:

1. It uses a microprocessor which is compatible with the Intel, IBM or AMD designs. Current examples are the Intel Pentium 4 or Celeron, and the AMD Athlon, Duron and Thunderbird.

2. It uses a program called MS-DOS or PC-DOS as a master controlling system (an *operating system*), to enable it to load and run all other programs.

3. It can be expanded, meaning that its facilities can be increased, by plugging in additional circuit boards into slots provided on the main motherboard.

4. It maintains compatibility with the previous PC types of computer.

The IBM PC was announced in November 1981, about 16 months after work had started on the design of the machine. The Intel 8088 microprocessor was used along with a set of support chips from Intel, and even by the standards of 1981 the specification of the machine was not particularly impressive. IBM did not initially see the PC as a business machine; even the name Personal Computer indicated that the intended market was the home user, and the price ensured that only the US home user could afford the machine.

These first PC computers were not of a particularly advanced design even for 1981; they had a small memory size and no magnetic disks were used – storage depended on connecting a cassette recorder. In 1984 the machine was re-designed and re-launched as the PC/AT, illustrated here, with much more memory, one or two magnetic disk drives and a hard disk option, and optional graphics display boards that could make use of either a colour monitor or a high-resolution monochrome monitor.

This is the form of the PC machine that was to become the standard for business use from the 1980s onward, and which was so extensively copied that the word clone entered the computer user's vocabulary. From the time of the launch of the XT, disks made by other computers were dubbed as either compatible or non-compatible. The subsequent history of desktop computing has been of the gradual submergence of the non-compatible except where specialized markets could make an incompatible machine viable. The PC may be manufactured by anyone; but incompatible machines are each the product of a single supplier.

- The format of the AT machine is now known as ISA, meaning industry standard architecture. This is by now almost obsolete, and modern fast machines owe nothing to the ISA type of design and construction.

NOTE: Computers are generally replaced at frequent intervals by more modern designs, so that you are less likely to be required to service old models. Servicing an old model could easily cost more than replacing it.

System block

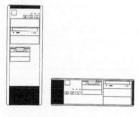

A modern microcomputer system consists of three main units, the system (or CPU) unit, which can be a desktop box or tower casing, the monitor, and the keyboard. To this we would normally connect other units called peripherals. The peripherals are units that need additional programs, called drivers, to operate, and they extend the use of the computer. Typical input peripherals are the mouse and scanner; typical output peripherals are the printer, loudspeakers and external drives. There are also peripherals such as the modem which provide for both input and output. The modern type of casing is distinguished by the letters ATX, a re-design of the original AT type.

Peripherals are normally connected through port connectors on the main system unit. A *port* is a complex circuit that interfaces the high data speeds of the computer to the lower speeds of other units, and also changes data voltage levels if required. In the past, these port connectors have been specialised, one type used for the printer, one for the modem and separate connectors for keyboard and mouse. More recently, a form of connection, the universal serial bus (USB) has been devised that allows a wide variety of peripherals to be connected to any of the sockets on the system unit. A particular feature of USB is that peripherals can be *hot-plugged*, meaning that they can be plugged or unplugged while the computer is running.

A simplified block diagram of the PC type of digital computer is shown, left. The system unit (or CPU) is the main processing unit in a desktop or tower case, and is connected to the monitor and the keyboard. The keyboard is the main input unit, and the monitor is the main output unit. The programs that are read from storage into the memory of the system unit will control the processing of inputs into outputs. As a simple example, the action of pressing the A key on the keyboard will generate binary codes

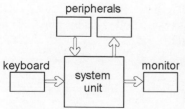

1 kilobyte (Kb) = 1024 bytes
1 megabyte (Mb) = 1024 Kb
1 gigabyte (Gb) = 1024 Mb

Hardware, software and firmware

that the system unit converts into signals that can produce an image of the letter A on the monitor screen, and a code for the letter A can also be stored in the memory.

The data unit is the *byte*, a set of eight bits. This is also the unit that is sufficient to code one letter of the alphabet. The larger units listed here, left, are also used – note that the ratio of sizes is 1024 rather than 1000, because 1024 is 2^{10}.

These terms are used frequently in computing. *Hardware* means the physical devices such as disk-drives, printers, computers and even the integrated circuits from which a computer is constructed. *Software* consists of the physically intangible items like the programs without which the system cannot operate. If the software programs are stored in a hardware (non-volatile) device such as a ROM chip or a CD-ROM disc, this composite item is then referred to as *firmware*.

Question 20.1

Name the three main blocks of the PC computer. What name is used to class items such as mouse, loudspeakers, printer, etc.?

System software

A computer system is totally useless without the programs that we collectively call *software*. Software consists of a collection of bytes that form instructions and data for the microprocessor in the system unit. In order to run a program, the bytes of software must be placed into the memory of the computer, and the location of the start of that memory block notified to the microprocessor. This is typically done by copying the 32-bit address of the start of the memory block, along with a number to show the number of bytes of the program, into locations (registers) inside the microprocessor.

One form of system software is described as BIOS, meaning basic input/output system. This is a comparatively short set of programs that allow the microprocessor unit to read from the keyboard and the disk drive(s), and to provide outputs to the monitor. This software is permanently held in the form of a chip, the BIOS chip so that it is always available when the computer is switched on.

The other important piece of system software is the operating system (OS). The older PC machines used MS-DOS, the Microsoft disk operating system, which is compact enough to need very little of the memory, but which can carry out only one action at a time. In addition, using DOS requires the user to type in a command word, followed by pressing the ENTER/RETURN key, for each command.

Each program can create documents, items of data. Typical documents are text (from a word processor or editor), worksheets (from a spreadsheet program) and drawings (from a graphics program).

Later machines have used the Windows operating system. This makes use of MS-DOS, but in a way that is easier to learn and use. The commands are selected by using the mouse to move a pointer on the screen, and clicking a button on the mouse when the pointer is over a command that is usually one of a set in a menu. Clicking on the name of a program will run that program, and clicking on the name of a document will run the program that created the document (if available) with the document loaded for editing.

Activity 20.1

Start up (boot) a PC machine that will run Windows. Use the mouse to select a program and run the program. Use the mouse to select a data item and delete it. Use the mouse to select a set of data items and copy or move them to another folder.

The motherboard

The motherboard is the main printed circuit board of the PC computer, usually consisting of several layers of circuitry. The motherboard contains a socket for the microprocessor, along with sockets for memory, power supplies, input/output connectors and expansion cards.

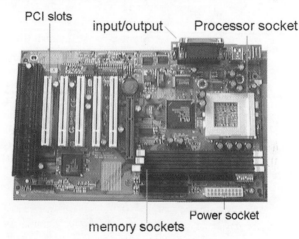

Figure 20.1 A typical modern motherboard with some important parts labelled.

PC technology changes rapidly, and any photo of a motherboard will be out of date in a year or so, but the general layout is fairly constant. In the illustration, the older type of microprocessor socket is shown, but the more modern types differ only in the number of pin-holes. Modern motherboards are predominantly of the ATX type, intended to be used in ATX cases.

The microprocessor socket is always of a ZIF variety, meaning zero insertion force. A lever is used to release clamps around the pinholes so that a microprocessor can be dropped into place. The lever is then moved to its lock position so that the chip is securely held. Modern sockets use 423 to 630 pins, depending on the processor type, but for some time it was more fashionable to use a slot-fitting, with the microprocessor soldered to a small board along with memory chips (the cache, see later) that fitted into a slot.

A battery, once a nickel-hydride type but now more likely to be a lithium cell, is clamped to the motherboard. This provides power backup for a small portion of memory, called CMOS-RAM, that holds information on the setup of the computer (such as the hard drive values). If this memory is cleared by disconnecting the battery, the computer cannot be used until the values are either typed in or found by starting an automatic scan of equipment.

The expansion slots are shown as PCI (PC interface), replacing the older ISA (industry standard architecture). Modern motherboards also include a slot that can cope with bus signals at higher speeds (typically 66 MHz or more). This AGP (advanced graphics processing) slot is used for the video graphics board that interfaces the computer to the monitor. The memory slots illustrated are for the DIMM type of memory boards (see later). Slots for faster types of memory are becoming available at the time of writing.

At one time, connections for such items as input/output and interfaces for the floppy disk and the hard disk were made using plug-in cards, but nowadays these features, and the CD interface, are included as part of the motherboard. Some motherboards also include the interfaces for the monitor and for sound output, but it is still common to use separate cards for these actions. The sound card fits into a PCI slot, and the graphics card into either a PCI or an AGP slot, depending on the card design. At the time of writing, most graphics cards used the AGP fitting. The serial and parallel interfaces, along with sockets for keyboard and mouse, are all built into the motherboard. The trend at the time of writing is to use a single type of connection, the USB (universal serial bus), see later, for all connections.

Computer technology is fast-moving, so that descriptions of motherboards and other components soon become out of date. In general, both speed and capacity (memory and drives) increase continually.

Power supplies

The power supply units (PSU) for PC computers are contained in a sealed box inside the main system case, with connections to the main switch on the casing. The power supply has in the past been a 200 W unit, but the fastest modern computers (1 GHz and above) require a 300 W unit. This is a switch-mode form of power supply, which makes servicing difficult, but the cost of replacement power supply units is so low that repair of a faulty unit is seldom economical.

Typical supply voltage and current levels are:

| +5 V | 20–25 A | –5 V | 0.5 A |
| +12 V | 4–7 A | –12 V | 7 A |

Note that the current output of the +5 V supply is very high, so that care must be taken against short-circuiting this supply. Most PSU designs incorporate fold-back protection, meaning that the output voltage will be reduced to almost zero in the event of a short-circuit.

Microprocessor

This block contains three important sections:

- A control section to generate all the necessary timing functions.

- An arithmetic and logic unit (ALU) to carry out the required calculations in either binary arithmetic or binary logic form.

- An accumulator (register) to hold the results of any operation before it becomes convenient to pass them to the memory, either for storage or for output.

The action of the microprocessor is to read in one or more bytes of instructions (called *opcodes*) from the memory, and to carry out these instructions. Typically this will require some data to be read from the memory, and the microprocessor will then carry out logic actions on the data and store any results in the memory. At one time many microprocessor actions required no more than reading data from one part of the memory and storing the same data in another part of the memory. Actions of this type are now carried out by another chip, the DMA (direct memory access) chip rather than by the microprocessor.

All the actions of a modern computer are the result of logic actions carried out one at a time by the microprocessor under the control of the program. The program must contain the command bytes for each instruction together with bytes that provide a location in memory for reading or writing data.

The capabilities of a modern computer are largely due to its high speed. The microprocessor itself is governed by clock pulses, generated from a pulse generator circuit that is part of the microprocessor chip. Whereas early microprocessors used clock rates of around 4 MHz, modern types can run at much higher clock rates, typically in the 1 GHz region.

The microprocessor clock rate determines how quickly the actions of the processor can run, but the overall speed of a computer depends also on other factors, such as:

- How quickly data can be read from or written to memory, typically around 133 MHz.

- How quickly signals can be passed to the other cards on the motherboard, typically 66 MHz.

- How quickly signals can be read from or written to a magnetic disk, typically 15–66 MHz.

- How fast the data can be transformed into a display on the monitor, typically 66 MHz.

Note: These speeds have been quoted in MHz rather than the more usual MB/s (megabytes per second) to show the typical frequencies that can be encountered.

The speed of a particular set of instructions depends on what percentage of the actions are of the slower type, so that computers are often compared by using *benchmarks*, meaning programs that contain a large number of actions of one particular type, testing each of these different speeds.

Early computers used the byte of 8-bits as the unit of data, and microprocessors of that type could work with one byte at a time. Since then, the capabilities of microprocessors and computers have steadily expanded, and units of 2, 3, or 4 bytes are more common. A data unit of more than one byte is called a word, and the usual word size for a modern PC is a 4-byte word (32 bits). This is likely to increase to 8 bytes (64 bits) in the lifetime of this book. The advantage of using a larger word is that more processing can be carried out in each microprocessor clock time.

Question 20.2

Name three factors that contribute to the high speed of a modern PC computer.

Memory

In addition to containing the program that controls the way in which the computer operates, the memory has to hold or store the data items that are waiting to be processed. This requires two types of memory device. One is called a *read only memory* or ROM. The ROM will retain its information even when the power is switched off and it carries the most basic operating instructions. The second has to be capable of being written to and read from is known as a *random access memory* or RAM.

Because of its retentive memory, the ROM is described as being non-volatile, while the RAM which loses the data when the power is switched off is said to be *volatile*. Memory is made up of MOS IC chips, and it can be read or written in very short times, typically 70 ns. The size of RAM for a modern PC is typically 64 Mb or more. While the computer is being used, data is being read from the hard drive into the memory, and if the data is changed during the use of the computer it must be written back to the hard drive in its altered condition.

The ROM chip that is needed to provide programs that enable the computer to operate in the most basic way (using disk drives, monitor and keyboard) is referred to as the **BIOS**, meaning Basic Input Output System.

When a PC machine is switched on, program instructions are read from the BIOS, and these are used to load in the main operating system (such as Windows) from the hard drive. During this time, the instructions in the BIOS are usually copied to the RAM because RAM is much faster than the ROM used for BIOS.

Another type of memory that is used in modern PC machines is *cache* memory. Cache memory is volatile but very fast, and its use considerably increases the operating speed of the PC. The principle is to read a block of data from the hard drive, and then allow the microprocessor to read from or write to this cache of data. This works because reading or writing a block of data (typically 128 Mbytes) is much faster than carrying out individual read/write actions on the same amount of data on the hard drive. The microprocessor can then read from or write to the cache at a much higher speed. Sometimes the data that the microprocessor needs is not present in the cache, and another block must be read, but with good program planning, some 90% of the data that is being used at a particular time will be held in the cache. Modern computers use more than one cache, with a very fast cache (the primary cache) built in as part of the microprocessor chip, and another (the secondary cache) obtained by using part of the main RAM memory.

Answers to questions

20.1   System unit, keyboard and monitor. The others are peripherals.

20.2   Fast clock rate for processor, fast disk read and write, fast memory read and write.

21

Data storage devices

**The disk units –
hard drive**

These devices act as an extension of, or backup to, the working of the computer's internal memory. During normal working, the computer may need to perform a variety of tasks, each requiring a different program. These service programs, as they are called, are conveniently stored on magnetic disks. Unlike the RAM of the computer, data stored on magnetic disks is *non-volatile*, so that it is available whenever the computer is switched on. Magnetic recording of data is not like analogue tape recording, and several systems are in use. All, however, rely on saturation recording, meaning that the magnetic material is magnetized to the maximum extent possible, with the direction of magnetism used to indicate a 1 or a 0 bit. The systems that are now used are more elaborate, but the principles are the same.

Disk (or tape) units are also valuable for storing masses of data of the type that might be gathered during such applications as process control or quality control. Since both these forms of backup devices handle binary data in a way that is not normally acceptable to the computer as it stands, they also need interfacing in the same way as did the input.

The main type of disk for a modern PC type of computer is the hard drive. This consists of many magnetic discs (or *platters*) coated with a magnetic material, all revolving on the same shaft. Electromagnetic heads, similar in principle to tape heads but much smaller, can be placed at any point on each platter to read or write data; all the heads are moved together, not independently. The head movement is controlled by a voice-coil mechanism, similar in construction to the movement of a moving-coil loudspeaker, controlled in a feedback system. All this is controlled by the microprocessor. The rotational speed of a typical hard drive is 7200 rpm, with some types using 10 000 rpm or more. Such speeds are possible because the record/replay heads are not in contact with the platter surfaces, but float on a thin film of air, like a hovercraft. The whole mechanical assembly is contained in an air-tight casing that has been assembled in a 'clean room' to avoid the entry of dust.

WARNING: The casing of a hard drive must never be opened. Nothing inside the casing will be useable if the casing has been exposed to air containing dust and smoke.

- A typical hard drive will have a capacity of 5 Gb or more, and access times are in the order of a few milliseconds.

- Disk drive heads are manufactured using thin-film coils rather than wire wound coils. Increasingly, magnetoresistive heads are being used for reading because they provide larger signal outputs. A magnetoresistive material is one whose resistance changes when it is affected by a magnetic field, and such materials can be used in very small sensors manufactured using the same type of techniques as for ICs.

Figure 21.1 shows a block diagram for a hard drive, omitting power supplies. The data cable connector supplies signals that control the drive motor and the position of the read/write heads, and the data signals are passed to and from the heads.

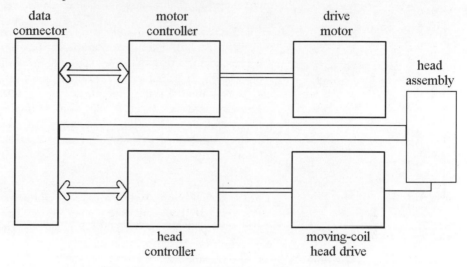

Figure 21.1 Block diagram of hard drive, simplified

Activity 21.1

On a motherboard, identify the following items:

(a) CPU, (b) memory, (c) video graphics card or interface, (d) ports, (e) battery, (f) expansion slots.

Drive controllers

Older computers used a set of chips on the mainboard as a drive controller. The action of a drive controller is to change the binary code for writing into a format that can be recorded magnetically, and perform the reverse of this set of actions for reading. This interface is needed because the microprocessor can operate at a much higher speed than can be used for recording or reading, and because a magnetic medium cannot record digital signals directly. Modern hard drives carry the interfacing circuits as part of the drive rather than as a separate card inside the computer. This system is referred to as IDE, meaning integrated drive electronics, and the version currently in use is EIDE, with E meaning extended. An older system, SCSI (small computer systems interface) is still in use for machines that require a large number of drives.

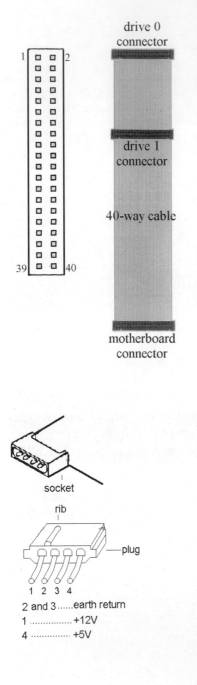

drive 0 connector

drive 1 connector

40-way cable

motherboard connector

socket

rib

plug

1 2 3 4

2 and 3 ......earth return
1 .... +12V
4 .... +5V

In addition, the higher speed of modern computers makes it undesirable for the microprocessor to control hard drive reading or writing. This is now done using a separate chip, the DMA (direct memory access) chip. The main microprocessor will pass to the DMA chip the details of where the data is located, where it has to be copied, and the number of bytes of data, and the DMA chip will then carry out the transfer, leaving the microprocessor free to work on another action.

The EIDE system is virtually a standard on PC machines other than those used as servers for a network, and the hard drive is connected by way of a 40-way cable. At one end, a socket engages with the EIDE socket on the motherboard, and the other two sockets are used for hard drives, with the end socket used for the drive that will be lettered C. On modern PC machines, there are two EIDE plugs on the motherboard, referred to as the primary and secondary EIDE connectors. This allows up to four devices to be connected, typically two hard drives, a fast CD-ROM drive and a slower CD writing drive.

The cables that are used are of the ribbon type, and the connection to Pin 1 of the sockets is marked by a red stripe or some similar identification on the end wire. When cables are connected up, care must be taken to ensure that this striped side of the cable is at the Pin 1 end of the connector. If a connector is reversed the drive will not be correctly connected and it will not be under the control of the computer.

In addition to the data cable, each drive needs to use a power cable. These are taken from the PSU in the casing, and typically six will be provided to be used on hard drives, a floppy drive, a CD-ROM drive and any other drives that are needed, such as CD writer or tape backup drives. The power plug uses a 4-way cable, and the usual form of plug and socket is illustrated, left. The rib on the socket and the corresponding groove on the plug ensure that the plug cannot be inserted incorrectly. A miniature version is sometimes used for 3½ inch floppy drives.

Before a new hard drive can be used it must be formatted, meaning that a set of tracks on each platter is magnetically marked into tracks and sectors with an identifying byte, and the position of the start of each sector is recorded in one of the tracks. A track is a circle on a platter, and a sector is a portion of a track, conventionally holding 512 bytes. The identifying portion is called the *directory track* or *table-of-contents* (*TOC*); its use allows data to be split up and recorded on different tracks and sectors, with the directory track information used to specify where each portion of data is placed.

To format a hard drive, a program must be used. Since Windows cannot be installed on a hard drive until the drive has been formatted, the program runs using MS-DOS, read from a floppy disk. The program is called FORMAT.COM, and with MS-DOS floppy in the drive and in use, formatting a hard drive is carried out by typing the command:

FORMAT C:

assuming that the hard drive uses the letter C, as is normal. Formatting will take several minutes, with the exact time depending on the size of the hard drive and the speed of processing. Once the drive is formatted, the Windows operating system can be installed by using the CD-ROM drive, which can be provided with a driver from the same floppy disk that contained the MS-DOS system files.

The usual hard drive is fixed, and the main drive, lettered C, is used to hold the operating system. For a small computer, this drive may also be used for programs and data also, but it is becoming more common to add a second hard drive for data. A removable hard drive can also be used for backup. The hard drive fits into a casing that is connected to the computer through the parallel port or by way of USB or Firewire ports, usually with a separate power supply. Backup files can be copied to this drive, and the drive then removed from its casing and kept in a safe place. This type of backup combines fast copying with ease of use, and can be used, for example, to ensure that all the data that has been altered in a day is backed up and put in a safe each evening.

Question 21.1

Describe the connections that have to be made to a new hard drive. What action is needed before a new drive can be used, and how is this carried out?

Activity 21.2

Using a multimeter, check connections on (a) an IDE drive cable, (b) a parallel printer cable.

Hard drive problems

Hard drives are mechanical components that are continually running while a PC is active, so that they have a limited life. Inevitably, a hard drive will fail, though this may be after many years use, and the computer may have been scrapped and all the data transferred before a drive failure occurs. The most serious failure is one that affects the main drive, because this makes it impossible to load the operating system. If a separate data drive fails it cannot be used, but the computer remains useable.

Typical failure problems concern either electromechanical components or disk corruption. The main drive motor can fail, making it impossible to use the drive, so that the operating system will not load. This type of failure is easy to diagnose because you will not hear the drive motor start when the computer is switched on. Another form of electromechanical failure affects the voice-coil drive for the heads, so that the tracks cannot be correctly located. In this type of failure also the operating system will not load, and the usual clicking noise of the head system will be absent.

Failure of the electromechanical system will require the drive to be replaced, but the data that was recorded on the drive will still be intact. Specialist firms can open the drive in a clean room and copy the data to a backup drive so that you can restore your data. Remember that the value of the data on a computer system may be very much greater than the value of the computer itself.

Drive failure that causes corruption is much more serious, particularly on a disk that contains data. A typical cause is that a motor fault slows the platters down and the 'flying' heads scrape against the platter surfaces, tearing the magnetic material. It is very difficult, or impossible, to recover data from this type of head crash.

Because your data is precious, you cannot assume that a hard drive will be totally reliable, and you need to use *backups*. A backup is a copy of important data on a hard drive, and a good backup system will ensure that your data is copied to another medium at frequent intervals and can be easily copied back if a replacement hard drive is needed. Typical backup systems include floppy disks (1.4 Mbytes), tape drives (up to several Gbytes), CD (700 Mbytes) and DVD (up to 17 Gbytes) writing drives, and also removable hard drives (which run only during backup and are removed from a holder after backup is complete).

Activity 21.3

Fault-find and repair a machine on which the operating system fails to boot, and no sound can be heard from the hard drive.

Floppy drive

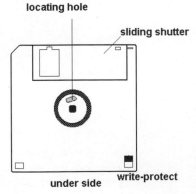

locating hole

sliding shutter

under side   write-protect

A modern PC computer will also use a floppy disk drive. At one time, twin floppy drives would be used, but this is now rare. The floppy disk drive allows the use of 3.5 inch magnetic disks with a capacity of around 1.4 Mb. These are a hangover from older systems, and will probably be superseded by other types of drive eventually. So far, however, the use of a floppy drive is essential because it allows a computer to be commissioned with a new unformatted hard drive. A system floppy, usually included with a copy of Windows for OEM (original equipment manufacturer) use, contains the MS-DOS command files along with drivers for CD-ROM drives, allowing you to format a hard drive, copy files to it, and make the CD-ROM drive available for installing Windows.

The floppy uses the same scheme of tracks and sectors as the hard drive, but the recording system is standardized because a floppy recorded on one PC must be readable on another (hard drives are usually fixed and used by one machine only, and the platters cannot be separated). The head mechanism uses a stepper-motor rather than a voice-coil. A new floppy must be formatted before use, and can be re-formatted to wipe it clear of all data. The formatting can be carried out from Windows, or you can switch to MS-DOS and use the command:

FORMAT A:

to carry out the formatting. This is not a fast operation, and when you format a floppy you may find that the computer cannot be simultaneously used for other actions.

Interfacing for floppy drives is, on modern computers, included as part of the BIOS, replacing several older types of interfacing that are now obsolete and which required a separate interface card. The connector for the floppy drives is on the motherboard, and uses a 34-way flat ribbon cable with the usual strip to identify the Pin 1 side. Cables usually allow for two floppy drives, but only one is normally used. The floppy drive also uses the same type of power connector as the hard drive, though a miniature version is sometimes used.

The floppy disk itself consists of a thin plastic disc coated with magnetic material on both sides, and held in a plastic casing to protect it. When the disk is inserted into the drive, read/write heads are lightly clamped against the plastic disc on each side (opposing each other), and the disc is spun to about 300 rpm. A tab at the corner of the disk housing can be positioned so that the disk is write-protected – this is checked by the drive when a floppy disk is inserted and a signal prevents writing, though the disk can be read. Some computers will try to read from a floppy at switch-on time if one has been left inserted in the drive, and will deliver an error message because the operating system is not on the floppy. More usually, the PC can be configured to ignore a floppy when the machine is booted, so that the operating system is only ever read from the hard drive. This can cause problems if the hard drive fails, so that you need to know how to reverse this setting (in the CMOS-RAM) if you are using a floppy to install MS-DOS so as to diagnose a hard drive failure or format a new drive.

A floppy disk is much less well protected than a hard drive, and floppies need to be cared for. They should be stored in a cool dry place, preferably in a box designed for the purpose, well away from magnetic or strong electrostatic fields. Given such care, floppies can have a very long useful life, but carelessness can easily make the tracks on a floppy unreadable.

Question 21.2

All the following devices can be used for backup. Place them in ascending order of typical capacity and state typical values of capacity.

(a) CD-ROM, (b) floppy drive, (c) removable hard drive.

CD ROM drive

The CD-ROM drive allows the use of CD discs that carry computer codes, and these are extensively used to contain programs and other information (such as multimedia text). Older CD-ROM drives are read-only, but a later generation of drives permit both reading and writing. The conventional CD-ROM disc will hold about 700 Mbyte of data.

The CD system makes use of optical recording, using a beam of light from a miniature semiconductor laser. Such a beam is of low power, a matter of milliwatts, but the focus of the beam can be to a very small point, about 0.6 µm in diameter – for comparison, a human hair is around 50 µm in diameter. The beam can be used to form pits in a flat surface, using a depth which is also very small, of the order of 0.1 µm. If no beam strikes the disc, then no pit is formed, so that we have here a system that can digitally code pulses into the form of pit or no-pit. These pits on the master disc are converted to pits of the same scale on the copies. The pits/dimples are of such a small size that the tracks of the CD can be much closer – about 60 CD tracks take up the same width as one vinyl LP track. The pits are arranged into a spiral track starting near the hub of the disc and ending near the edge. Digital signals of 8 bits are converted into 14-bit signals to allow for error-reduction methods to be used, and also to avoid long runs of the same digit (0 or 1) occurring.

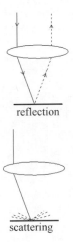

reflection

scattering

Reading a set of pits on a disc also makes use of a semiconductor laser, but of much lower power since it need not vaporize the material. The reading beam will be reflected from the disc where no pit exists, but scattered where there is a pit. The rotational speed of the disc is controlled so that the rate of reading pits is constant. This means that the disc spins faster at the start of the track than it does at the end. By using a lens system that allows the light to travel in both directions to and from the disc surface it is possible to focus a reflected beam on to a photodiode, and pick up a signal when the beam is reflected from the disc, with no signal when the beam falls on to a pit. The output from this diode is the digital signal that will be amplified and then processed eventually into an audio signal. Only light from a laser source can fulfil the requirements of being perfectly *monochromatic* (one single frequency) and *coherent* (no breaks in the wave-train) so as to permit focusing to such a fine spot.

- Focusing and track location are automatic. The returned reading beam is split and the side beams detected to that servo circuits can locate the track and focus on it. This mechanism must not be tampered with.

The CD reader drive uses the laser beam reading system that was developed for music CDs, but the speed of rotation of the disc is usually much higher. When you buy a CD-ROM drive, you will see this speed quoted as a multiple of the normal speed of the music CD, such as ×20, ×32 and so on. This allows the discs to be read at higher speeds, so that information can be read from a CD almost as fast as it could be from old hard drives.

The form of CD-ROM drives is fairly well standardized now, as a slim 5¼-inch casing. The front panel holds a drive-on light, a volume control and a headphone jack. This allows the use of the drive for playing ordinary

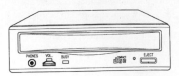

audio compact discs. The output audio signal is typically 0.6 V rms at 1 kHz.

When a button on the front panel is pressed, a tray slides out so that a CD can be placed in the tray, and pressing the button again will cause the tray to slide back in. The drive motor then spins the disc (taking a second or so to build up to full speed), and from then on the disc can be accessed, usually as drive D:. Windows allows you to specify *Autostart*, so that when a CD is inserted it will run its program(s) automatically.

- There are three connectors at the rear of the CD-ROM drive, the power connector feeding +5 V and +12 V DC, an audio output connector for amplifiers, and the data interface. The audio interface typically uses a 4-pin connector. The data interface uses a 40-pin IDC (insulation displacement connector) type.

The CD-ROM drive contains a low-power laser, but because the beam can focus to such a small point, it can cause damage to your eyes. No attempt should be made to use a CD reader with the casing removed, and if focusing problems are encountered, the unit should be returned to the manufacturer.

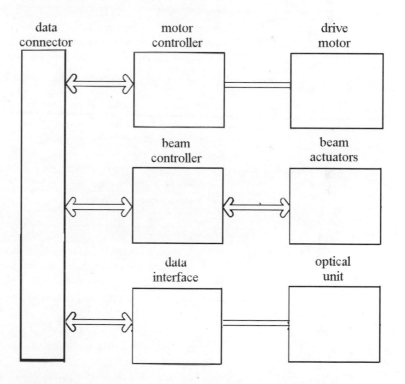

Figure 21.2   Block diagram, simplified, for a CD-ROM drive

Figure 21.2 shows a simplified block diagram for a typical CD-ROM unit. The control signals set the speed of the drive motor, which is not constant, it runs faster when the inner tracks are being read and slower for the outer tracks, keeping the data speed constant. The beam position drives select the track and maintain focus, and the optical head reads the signals reflected from the track. These signals are then converted into conventional digital data in the interface.

Question 21.3

Name three factors that make a CD-ROM ideal for backup purposes.

CD-R, CD-RW, and DVD

The fast CD-ROM drives are read-only, but recently both Write-once (CD-R) and Write-often (CD-RW) drives are now available at comparatively low prices. These drives use the same size and style of casing, but the read speeds of early types much slower, typically ×4 to ×6, with slower writing speeds, usually ×2. The write-once CD-R drives can use blank discs that are very attractively priced (at around £0.75 or less) and which like the standard CD will hold about 650–700 Mbyte of data. Compression can increase this to around 1 Gbyte. The CD-RW drives cost more, and the blank discs currently retail at around £2. Later models permit much faster reading and writing speeds, typically ×36 for reading and ×16 for writing.

The DVD (digital versatile disc) is a development of the CD, and a DVD player will accept compact discs as currently used. The DVD has been developed to allow much higher capacity, up to 17 Gbyte, so that motion pictures can be placed on a single disc and much larger capacity backup is available for computing purposes. The name is derived from the fact that the disc can be used for computer data, music (17 hours or more), still or motion pictures or any other form of digital data. The DVD units can be obtained in read-only or in read-write forms and are likely to replace the CD units which are now used.

Connections

The PC computer needs to provide for connections to and from peripheral devices such as printers, scanners, modems and other units. Until recently, keyboard and mouse connections were made using small sockets of the DIN type, but more recently the USB (universal serial bus) type of connectors have come into favour. Keyboard and mouse connections will be considered later.

Until recently, the main I/O (input and output) port connectors for the PC were either parallel (Centronics) or serial (RS232 or RS423). The parallel port transmits eight data bits at a time, and is used for the printer, since printers predominantly use this type of connection. The data cable for a parallel port requires more than eight lines, because others are used for synchronization. Typically, seventeen lines are used for signals, and the other lines are used for earth returns.

The serial ports were at one time used for modem connection, but the extensive use of internal modems (as a PCI card) has reduced the need for this. The serial port transmits one bit at a time, and is still in use as a mouse connection if no dedicated mouse port if supplied on a motherboard. More recently, all of these types of connection are being phased out in favour of the USB (universal serial bus), particularly now that the faster USB-2 is available.

Parallel port

The older style of parallel port is often termed a Centronics port because its standardized format was due to the printer manufacturers Centronics who devised this form of port in the 1970s. Centronics port sockets on most small computers use a 36-pin Amphenol type of connector, but the PC uses a 25-pin D-type female connector at the PC end of the cable, and the 36-pin Amphenol type at the printer end. This can be confusing, because the older serial port (COM1) connector on the PC is the 25-pin D-type male connector.

The original form of Centronics parallel port was intended for passing signals in one direction only, from a computer to a printer. Several designers made use of the unmodified system for bi-directional (two-way) signals by using the four signal lines that communicated in the reverse direction along with four data lines so as to get 4-bit bi-directional signalling. This in turn gave rise to a standardized system for allowing the use of the parallel port for 8-bit bi-directional signalling. This is the standard IEEE Std.1284-1994 system, and is otherwise known as the EPP (extended parallel port) system. A version of this, ECP (extended capability port) is now a standard fitting for PC machines. The speed of data can be 50 to 100 times faster than was possible using the older Centronics port, but the ECP connection on a modern PC is still fully compatible with older printers and other peripherals that use the parallel port.

Three types of connectors can be used. One, Type A, is the existing DB25 type updated to 1284 electrical standards. Type B and C are 36-pin connectors, of which the Type C is the standard that is recommended for new designs. Type C is smaller than older 36-pin types, uses a simple clip as anchor, and permits the use of additional signals, *peripheral logic high* and *host logic high*. These additional signals are used to find if the devices at each end of the cable are switched on. The cable structure uses twisted pairs of signal and earth return wires, all surrounded by an earthed screen. Ribbon cable can be used on connections between the motherboard and the port sockets on the computer casing.

- Where ribbon cable is used, the line corresponding to Pin 1 is marked in some way, usually with a red stripe.

Serial ports

Serial ports are used for modems, for serial mice, and for linking PC computers together into simple networks. These ports send or receive one bit at a time, and at their simplest they need only a single connection (and earth return) between the devices that are connected. Serial ports are seldom quite so simple, however.

- The keyboard and PS/2 mouse use serial ports, but these are dedicated types, meaning that you cannot use them for other devices.

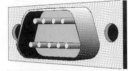

The serial ports we are dealing with here are the COM types used for connecting other peripherals.

The serial transfer of data makes use of only one line (plus a ground return) for data, with the data being transmitted one bit at a time at a strictly controlled rate. The standard system is known as RS-232, and it has been in use for a considerable time with machines such as teleprinters, so that a lot of features of RS-232 seem pointless when you are working with modern equipment.

The original cable specification of RS-232 was for a connecting cable of 25 leads, as shown in the illustration. Many of these connections reflect the use of old-fashioned telephone equipment and teleprinters, and very few applications of serial cables now make use of more than eight lines. In addition, the original RS-232 specification provided for signals of +15 V and –15 V.

A more modern specification, RS423, is now used, with fewer lines and signal voltages of +5V and 0 V in line with the digital voltages levels on the motherboard. The standard connector for PC machines is now the D-type 9 pin, illustrated 3, but even in this respect standards are widely ignored and some manufacturers use quite different connectors.

If cables are supplied along with serial equipment then you have a better chance of getting things working than if you try to marry up a new piece of equipment with a cable that has been taken from something else. The important point is that you cannot go into a shop and ask for an RS232 or RS423 cable, because like the canned foods, serial cables exist in 57 varieties. When you need a cable, you must specify precisely what you want to connect with it.

The advantages of using serial connections, however, outweigh the problems, because when a modem is connected by a serial cable to your computer you can use a simple single line (telephone or radio link), and distance is no problem – wherever you can telephone or send radio messages you can transfer computer data providing that both transmitter and receiver operated to the same standards. A huge variety of adapters (such as *gender-changers*) can be bought to ensure that your cable can be fitted to a socket that may not be of the correct variety, but this does not guarantee that the connections will be right.

- RS232 or RS423 serial links can be used up to their maximum speed of 115 000 bits per second, but no faster. This is one of the drawbacks of serial ports that has led to the development of faster serial systems such as USB and Firewire.

The serial ports are referred to as COM1, COM2 and so on, but the use of more than two serial ports on older motherboards was fraught with problems because a COM3 port, for example, had to share the IRQ4 interrupt signal with the COM1 port. This is not a problem for modern motherboards, and it's quite normal for an internal modem to use its own COM3 port without interfering with the actions of the COM1 and COM2 ports.

The universal serial bus (USB)

The essential simplicity of serial connections for linking a computer to a peripheral has spurred designers into looking for something better than the old telecommunications serial ports that belong to the pre-computer age. What we need is a type of serial port that can be used for all the normal computer connections, and which can be connected in 'daisy-chain' type of network with each device connected to another with only one of them needing to be connected directly to the computer.

The answer to this is called the universal serial bus (USB). It has been designed to be *hot-plugged*, meaning that devices can be connected and disconnected with the computer switched on and working. This is possible only if the system is supported by the computer, the peripheral device and the operating system. At the time of writing, USB 1 is fitted to motherboards, but is being superseded by USB 2.0.

USB 1 permits communications between devices that are equipped with suitable interfaces at serial data rates ranging from 1.5 Mbits/s to 1.5 Mbytes/s. This is very much faster than the old-style serial port system. The inter-connecting cable can have a maximum length of 5 metres and consists of two twisted pair cables, one pair for power and the other for signalling. The distance can be extended to about 30 metres by using a *hub* terminal device as a line repeater.

Terminal devices, such as keyboard, mouse and printer, are added to the basic PC in a daisy chain fashion and each is identified by using a 7-bit address code. This allows up to 127 devices, in theory at least, to be connected. In practice, not all devices allow daisy chain connection (picture a mouse with two tails!), and so the computer needs more than one USB connector. In practice, the computer will be fitted with 2–6 USB connectors of the flat type, and each peripheral will use one connector of the square type, see illustration.

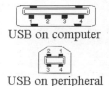

USB on computer

USB on peripheral

- Other terminal devices can include scanners, fax machines, telephone and ISDN lines, multi-media display and recording systems and industrial data acquisition devices.

- USB 2.0 has now been announced, running at 480 Mbps, some forty times faster than USB 1, and comparable in speed with Firewire (IEEE 1394). Hub units for USB 2 will work with peripherals that use USB 1, but USB 1 hubs will not work with USB 2 peripherals.

Question 21.4

A new computer has one parallel connector, two serial connectors, two USB connectors, and two PS/2 connectors. Suggest how these connectors might be used.

Cable problems

Cables are always a potential source of problems in PC equipment. It is rare for a cable to be faulty when it is unwrapped, but faults can develop, mainly due to excessive flexing of cables, when cables are fitted or when items of equipment are moved with the cables left connected. Where a cable connector provides for anchoring by a clip, this should be done, and anchoring is particularly important for printer cables, particularly for the 36-pin connector at the printer end. USB cables make no provision for anchoring, and care needs to be taken to avoid pulling out these small connectors when equipment is moved.

Cables can be checked by continuity testing or by substitution. The parallel type is most easily tested by substitution, and this is a preferred method because it involves the least amount of handling. If no substitute is available, you can check cables by checking for continuity, but this can be very time consuming, and you risk damaging a cable in the process of checking connections manually. A much more satisfactory method is to use automatic testing equipment which connects to each end of the cable and then carries out a set of continuity checks on each corresponding pair of pins.

- Note that data cables for computers are rated for low voltages only, and must not be tested using any type of unit that employs high voltages.

Activity 21.4

Install and commission a standalone (not networked) PC computer. Position the main unit (desktop or tower) and connect the monitor, keyboard, mouse, and printer. Switch on, and check that the computer boots into Windows. Use the Windows *Start* menu to switch off.

Answers to questions

21.1   Connect data cable and power cable. Formatting will be needed, using a floppy disk that contains the program for the FORMAT command.

21.2   The order is floppy drive, CD-ROM, removable hard drive.

21.3   It is read-only, removable, and has a large capacity.

21.4   Parallel port used for printer. Serial port for modem or for connecting to another computer. USB for scanner, printer, monitor, keyboard. PS/2 for keyboard and mouse.

22 Input and output devices

Keyboard

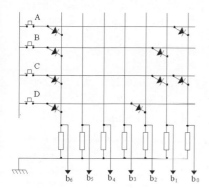

PS/2 type

The keyboard contains all the keys of a normal typewriter, together with a set of function keys (labelled F1 to F12) that can be used for controlling actions. These function keys are all programmable, meaning that their actions can be set by whatever program is running. A set of number keys is usually included at the right-hand side of the keyboard, and there are also keys marked with arrows that will move a pointer (or cursor) on the screen. In addition, there are keys marked Insert, Delete, Home, End, Page Up, and Page Down. All these key actions can be modified to suit whatever program is running.

A keyboard is constructed from a matrix of input and output lines, with the required interconnections being provided by suitably positioned diodes and contacts made by (de-bounced) switches. A section of such a structure is illustrated, left.

Before a key is pressed, each output bit line b_6 to b_0 is held at 0 volts or logic 0 by the action of the load resistors. When a key is pressed, + 5 volts is applied to the appropriate input line so that any diodes connected to it become forward biased and the corresponding output bit line is pulled up to + 5 volts or logic 1. In this way, each keypress generates a unique code that can be used by the computer and transmitted over a serial link. The link need not be fast, because the speed of using a keyboard is not fast. Some keyboards for laptop computers incorporate a pointing device (trackball) that substitutes for the use of a mouse on a desktop machine. The keyboard cable can be flat or round and conventionally uses a 4-way connection to a 5-pin DIN type plug or the later 6-pin PS/2 type, though USB keyboards (which include a separate USB mouse port) are now more common.

Keyboard maintenance should be confined to cleaning at a time when the computer is switched off. The keys become coated with a grimy layer from human fingertips, and this can be cleaned using a spirit-based cleaner. You should never use spray cleaners nor strong solvents. Keyboard problems are usually detected and notified when the computer is booted up, but the error message is often a number code which tells you nothing about the cause. Typically, PC machines use a three digit code in the range 300 to 399 to signify a keyboard error, and if you see this you need to switch off and check the keyboard connections and the keys themselves. The two most common causes of keyboard error are a loose plug or a jammed key. If in doubt, check by substituting another keyboard that is known to be perfect.

- Key action can be checked by starting a text application (such as Notepad) in Windows, and typing, using all the letter and number keys. If you have a program that makes use of the function keys, this also can be used.

Activity 22.1

Connect up a keyboard, stating what type it is (PS/2, serial, or USB). Boot the PC and check that the keyboard is working. Reboot with the A key held down to simulate the effect of a jammed jey. Carry out cleaning actions on the keyboard.

Mouse

The mouse is an alternative form of input which is used to move a pointer on the screen. This is useful for selection and for drawing actions. The mouse consists of a casing holding a heavy steel ball covered with a rubbery coating. The movement of the ball is sensed by rollers or by optical methods so that any movement of the mouse will generate two sets of signals, one for up-down and another for side-to-side movement. In addition, two or three switches can be operated by buttons on the mouse, and the action of pressing and releasing a switch (called *clicking*) can be used to send controlling signals to the PC.

- Some types of mouse also use a wheel that can be rotated to scroll the lines on the screen.

The mouse is an essential part of the GUI (graphical user interface) systems in which choices are made by moving the pointer and pressing (clicking) a button on the mouse. The best-known of these GUI systems are Microsoft Windows, used on PC machines, and the (older) operating system devised for Apple computers.

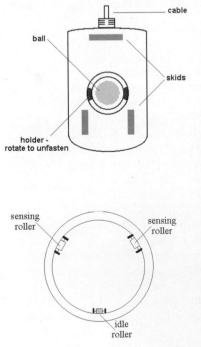

The illustration shows the underside of a typical mouse. The ball is held in a semi-spherical casing, and its movement is sensed by rollers that are held against the ball. One roller detects lateral movement, and the rotation of the roller shaft is digitized by using, typically, an optical system with a set of blades and an infra-red light-beam. Another roller detects movement in the direction at 90° using the same method. A third roller is a dummy, used to hold the ball against the other two. Circuits on a small PCB within the mouse shell convert the digitized signals into serial form for transmitting to the computer.

The mouse can be connected in three different ways, each of which requires a mouse with the correct cable termination, and a matching connector on the computer. Computers that use the ATX type of motherboard and casing will normally provide a mouse port with the PS/2 type of connector specifically intended for the PS/2 type mouse. Another option, used mainly on AT motherboards and cases is to fit the mouse cable with a 9-pin serial connector that is plugged into the serial port. The third option that is starting to appear is the USB fitting, plugging into a port on the (USB) keyboard. Adapters are available for DIN and PS/2 so that one variety of connector can be used with the other.

Connecting the mouse has no effect unless suitable mouse driver software is present and is run each time the computer is started. On modern machines that use Windows the mouse software is run automatically when

Windows starts. You can use the *Control Panel* action of Windows to modify some of the mouse actions to suit your own requirements. Older PC machines, using MS-DOS as the operating system, required the mouse driver software (a file called MOUSE.COM) to be loaded in each time the computer was switched on.

If you need to install a new mouse, or one that does not work with the default driver that Windows uses, the procedure is as follows (Windows 98/Me illustrated; others are similar):

Use *Control Panel* from *Windows Explorer* or from Start – Settings – Control Panel. Double-click on *Add New Hardware*. This starts the title panel of a Wizard, and you need to click the *Next* button. The next panel asks if you want to have Windows automatically detect your new hardware. This is the advised option, because if Windows detects the hardware it will almost certainly ensure that the correct software driver is installed. If you opt for manual detection then you will probably need a disc containing drivers from the supplier of the new hardware.

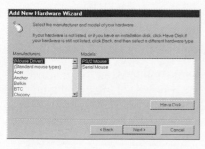

Before you start to use the auto-detect action, shut down all programs other than Explorer. When you start the auto-detect action, the progress is indicated by a bar-graph display, and the action may take several minutes. Do not worry if there is no change in the indicator for a minute or so, as long as you can hear your hard drive working. When the action is completed you will see a report on the new hardware that Windows has detected and on the driver that has been used.

If Windows does not detect a new hardware item, or if you opt for manual detection, you will see a list of manufacturers and equipment from which you can select. From the list that appears click on *Mouse* to see a list of manufacturers and models, left. If your type of mouse is not listed, then insert the disk that the mouse manufacturer has supplied, click the *Have Disk* button, and follow the instructions that appear on the screen.

Note that this method of detecting hardware and loading drivers applies also to other items.

Question 22.1

The 1982 IBM PC/AT computer did not use a mouse, but all modern computers do. What has changed to make the use of a mouse (or other pointing device) essential?

Mouse problems

The main problem you are likely to encounter with a mouse is sticky or erratic pointer movement, caused by dirt either in the mechanism or on the skids. Turn the mouse over and clean the skids with a moist cloth. If this removes the problems there is no need to do any more, but if the mouse still gives problems, remove the circular cover that holds the ball, revealing the rollers illustrated earlier.

- Dirt collects on the skids from the mouse mat, and this layer of dirt is enough to lift the mouse body so that the ball can no longer roll over the mat. If the skids are dirty, you should clean the mouse mat also.

Remove the ball and clean it, using a clean moist cloth — you can use spectacle lens cleaner fluid or windscreen-cleaning fluid if you prefer. Check by rotating the rollers with a finger that these are working correctly – rotating a roller should move the screen cursor when the computer is working. If the roller action is not working, you need a new mouse. If the rollers are working correctly you should clean them. These are most easily cleaned by wrapping a piece of clean cloth at the tips of tweezers, moistening this, and wiping it across each roller, wiping each roller several times while turning the roller slightly. Allow a few minutes for any moisture to dry and then reassemble the ball into the mouse casing and make sure that it is locked in place.

- If the mouse movement is erratic from the start, check that the correct driver is being used. For a machine using Windows and a standard type of mouse, it is most unlikely that the driver would be at fault, but if you have changed to another mouse or you are using a machine which has an unorthodox arrangement this is a possible cause of trouble.

- Another type of problem that you may encounter is that the mouse will move the pointer easily in one direction, but needs much more mouse movement to shift the pointer in the opposite direction. This cannot be cured by cleaning the mouse rollers (though you should try it), and usually requires you to buy another mouse. Test first by substituting another mouse.

The standard mouse uses two buttons, but there are some three-button types, though the third button is not normally used by Windows software. A more recent Microsoft mouse design, the Intellimouse, uses scroll wheels to permit the screen scrolling action to be carried out from the mouse. A scroll-wheel type of mouse is very useful if you work with long text documents.

Activity 22.2

Install a mouse, stating what type of connection is used. Boot the computer and when Windows has loaded check that the mouse is operating correctly. Clean the mouse, including ball and rollers.

There are other options for pointing devices, and one is the *trackball*. The trackball, illustrated left, looks like an inverted mouse, and consists of a heavy casing that remains in one place on the desk. This carries a ball, larger than the usual mouse type, that can be moved with the fingers, and a button either side of the ball. The action of ball movement and button clicking follows the same pattern as the use of the mouse, and some users prefer it, particularly for graphics applications.

Graphics tablets are another form of pointer that, as the name suggests, are particularly suited to graphics work. A graphics pad or tablet looks like a rectangle of plastic with a stylus, and the movement of the pointer on the monitor screen is controlled by the movement of the stylus over the graphics tablet. This means that every point on the monitor screen corresponds to a point on the graphics tablet, so that the larger the tablet you use the more precisely you can control the screen pointer. The stylus can be pressed to provide the click action of a mouse. Graphics tablets are expensive, particularly in the larger sizes.

Monitor

The monitor is a display device that is the main output for the computer, showing the effect of entering or manipulating data. At one time computers did not use monitors, and relied entirely on printed output. When monitors were first provided on large computers they were used only for checking that the computer was working correctly, hence the name.

NOTE: To avoid repeating information, you should refer to Unit 6 for details of TV displays. In particular, you should know the meanings for the terms hue, luminance, saturation, interlacing and convergence. You should be aware of the colour triangle for adding colours, and the structure and operation of the colour CRT.

For the lower-cost desktop machines, the CRT is the basis of the monitor. Colour monitors, whose resolution is always inferior to that of a monochrome monitor, were introduced in the mid-1980s, and are now standard. All colour monitors currently used for PC machines are of the RGB type, meaning that separate red, green and blue signals are sent from computer to monitor.

The **resolution** of a monitor measures its ability to show fine detail in a picture, and is usually quoted in terms of the number of dots that can be distinguished across the screen and down the screen – the figures differ because the screen is not square, with a ratio of width to height of 4:3 (as for pre-digital TV). These dots are referred to as pixels (picture elements). A figure of 640×480 pixels is now regarded as the minimum standard, the VGA (video graphics array) standard that was introduced in 1982. This resolution is considerably better than can be obtained on a domestic TV receiver. Resolutions of as high as 1280×1024 pixels (SVGA, with S meaning 'super') can be obtained on a comparatively low-priced colour monitor.

As you might expect, the signals within the computer cannot be used directly by the monitor, and the interface circuits are placed on a graphics card. The graphics card also contains memory, and will convert a set of digital signals into the analogue signals of red, green and blue that are needed by the monitor. For normal text work, the graphics card can be simple, but more elaborate cards are needed for pictures and particularly to handle animated pictures and video. Large amounts of memory (typically 8 Mbytes) are needed for fast-changing picture displays for video or games.

The monitor picture traditionally uses the ratio of 4 units of width to 3 of height; this is referred to as the aspect ratio. This is also the traditional TV aspect ratio, but new TV receivers are now following the wide-screen Euro aspect ratio of 16:9. Unlike TV which uses an interlaced picture format (see Unit 6), PC monitors always use non-interlaced scanning with high field (refresh) rates. As a result, the video bandwidth can be of the order of 80 MHz as compared to the 6 MHz of a TV receiver. Figure 22.1 is a simplified block diagram for a conventional CRT monitor.

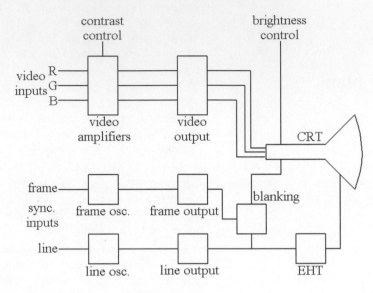

Figure 22.1   Simplified block diagram for a computer monitor using a cathode-ray tube

Monitors using CRT displays suffer from two forms of distortion, both of which can be corrected by adjustments. Barrel distortion makes the shape of a rectangle appear to have the longer sides curved, pin-cushion distortion causes the opposite effect. The corrections are made by modifying the shape of the scan waveform, and these corrections are nowadays usually digitally controlled by applying pulses to correction inputs in the IC that generates the scans.

Colour monitors have lower resolution than monochrome monitors because each element on the screen is made up of a red dot, a green dot and a blue dot. This makes the overall size of each element or pixel on the screen larger than the single dot of a monochrome display. The dot size of a colour monitor is often quoted as a measure of resolution. This is misleading, because the dot size has to be compared to the screen size – a dot size of 0.25 mm looks good on a 17" screen but would be hopelessly large on a 5" screen. You should look for the dots per screen size figure, such as 1024×768, as a guide to resolution.

Early monitors could cope only with signals that were of one fixed format, but NEC later developed monitor circuitry that could automatically detect the signal type, including frame rate and the type of synchronizing signals, so that a monitor could self-adapt to whatever signals were being sent to it. Such a monitor is a multisync type, and most modern monitors are of this form.

The conventional monitor, using a CRT, must now conform to the EnergyStar specification for reduced power consumption. For a 15-inch monitor, this requires a maximum power consumption of around 80 W, and provision for reducing this consumption in two stages to 15W (or less) and then to 8W or less. The lower standby consumption figures are achieved by switching off the scanning and the video output stages when the monitor has been idle for a preset time such as 15 minutes. In a large office, the use of power-saving can result in a considerable reduction in costs both directly and indirectly.

Activity 22.3

Using the manufacturer's service manual, carry out the recommended routine maintenance actions. Using a monitor known to be faulty, carry out diagnostic and repair actions. Clean the monitor, and check the Windows settings for resolution and number of colours.

LCD monitor

The LCD type of screen is an alternative to the CRT (cathode-ray tube) for monitor use, but at the time of writing is used mainly on laptops and on some higher-priced desktop machines. Larger LCD screens (15" or more) are prohibitively expensive for home users, but are increasingly found in business applications. A drawback of LCD monitors is that the picture is clearly visible only if you are directly in front of the screen.

Modern colour LCD screens use a huge number of LCD *cells* grouped into sets of three with colour filters between the cell and the (rear) light-source. The older type is described as passive, meaning that the cells are driven directly from an IC. Modern LCD colour screens are of the *active-matrix* type, using a separate (IC) electronic circuit controlling each cell, and it is this that makes such screens so expensive, accounting for the high cost of laptops as compared to desktop machines of the same capabilities. Modern screens are of the transflective type, which can be used either with backlighting (in the dark) or with frontal lighting (room lighting).

Question 22.2

When you change from 800 × 600 resolution to 640 × 480 on a CRT monitor, the picture remains full size but looks coarser. What happens when you make the same change on an LCD screen?

Other peripherals

Peripherals are devices that are not built into the main body of the computer, but which are connected externally. The keyboard and mouse are peripherals, and other important peripherals are the monitor and the printer. Many systems now include a *scanner* that can read documents, converting printed words on a document into computer data for a word-processor, and printed images into computer image files. Another form of scanner can be used to convert the images on slide or negative film into digital files.

Answers to questions

22.1   The mouse is essential to the use of GUI systems such as Windows

22.2   The picture becomes smaller because it is using fewer of the LCD cells.

Printers

Printers

Many types of printers have been marketed for computers, but the dominant varieties are the impact dot-matrix, the laser printer and the inkjet types. Virtually all of them use the standard Centronics connector or the later USB connector so that you can use any printer with any computer.

A matrix is a set of points arranged in rows and columns, and dot-matrix refers to making an image by using dots of light or ink in a matrix pattern. On that basis, the screen display is also a dot-matrix display, and both ink-jet and impact dot-matrix printers use the principle also. The printing head of a matrix printer can print a vertical line of dots, and the shape of each character is created by printing some dots in a line, shifting the head by a small distance, printing some other dots in a line, and so on until the complete character shape is printed. The resolution of the printer is determined by the number of dots in the line and the distance that the head can move in each step.

The impact dot-matrix was one of the earliest forms of printers for small computers. The paper is marked by being struck by a set of small pins that press an inked ribbon against the paper, leaving a dot mark for each pin. The first types that were considered acceptable for office use used a 9-pin head with a fairly coarse step, and the characters looked crude and 'dotty'. Modern 9-pin impact printers are available for heavy-duty, often at quite high prices. Later machines have used 15 or 24 pins in line and can print presentable characters and good-quality graphics images. Because of the dot structure, any character of any language can be printed, a considerable advantage compared to the limited range of the old daisywheel printers. The advantage of these machine for office use is that they can provide carbon copies, something that is not possible with ink-jet or laser printers.

Inkjet printer

The inkjet printer uses the same matrix principle, but instead of moving pins and a ribbon, a set of tiny jets is used to squirt a fast-drying ink at the paper. The two main systems currently are the bubblejet and the piezoelectric jet. The bubblejet system, devised by Canon, uses a tiny heating element in each jet tube, and an electric current through this element will momentarily vaporize the ink, creating a bubble that expels a drop of ink through the jet. The piezoelectric system, devised by Epson for its Stylus models, uses for part of the tube a piezoelectric crystal material which contracts when an electrical voltage is applied, so expelling a drop of ink from the jet. This latter system has been developed to allow for high-resolution printing of 720 dots per inch or more, and bubblejet models of high-resolution are also now available.

The ink is contained in one or more cartridges. Nearly all ink-jet printers manufactured now are colour types, and the usual cartridge scheme is to use one black-ink cartridge and one three-colour cartridge. Some models use a separate cartridge for each colour. Each cartridge is clipped securely in a cradle that ensures that the electrical contacts are firmly pressed

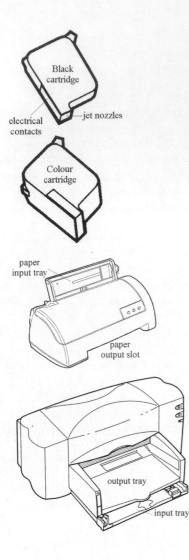

together and that the ink nozzles are correctly lined up. Printer software contains programs for lining up the colour cartridge(s) so that coloured lines match the positioning of black line. See later in this chapter for details of how to replace or install a cartridge. The ink is designed to dry quickly so that sheets do not smear when they are ejected from the printer.

The cartridge cradle can move from side to side on rails, controlled by a toothed belt that in turn is driven by a motor. The electrical connections are through a ribbon cable that is long enough to cope with the full range of movement of the cradle. The head drive motor is controlled by the printer software.

The paper is usually in single-sheet form fed from a hopper. Some printers place the hopper at the rear of the printer, feeding sheets down into the printing plane and then out at the front. This system relies partly on gravity for feeding, but the paper path is fairly straight, allowing thick sheets to be used. The illustration is of an Olivetti printer. Another method is to place the paper in a horizontal holder that is at the front of the printer, and pull each sheet through, bending the paper through 180° to the printing plane, then ejecting the paper on to a tray above the feed tray. This makes the printer more compact, but it cannot handle the thicker papers. The illustration of this type shows the Hewlett-Packard 895Cxi machine.

- The differences are important if you use coated inkjet paper that is coated on one side, or if you want to print both sides of a sheet. The rear-hopper system requires the coated side of the paper to face towards the printer, the front-holder method requires you to load the paper with the coated surface downwards.

Printing speed was at one time quoted in terms of text characters per second, and the figure referred to rather artificial conditions, so that it served only as a guide to the speed of printing that you might observe. Speeds are now more often quoted in pages per minute, and most printers can use more than one printing speed. The HP 895 series, for example, can print in draft (Econofast) mode at 10 pages per minute, in Normal mode at 5 pages per minute, and in Best mode at 3.8 pages per minute. These are black (text) printing speeds, and for most purposes the draft mode is perfectly acceptable. On some other types of machines, draft mode is too faint for most purposes.

Printing speeds will be much slower if a document contains graphics or elaborate text fonts, as compared to a document that uses a simple font throughout. Colour printing is always considerably slower, and an A4 colour page may need several minutes to print in *Best* mode.

- The font, or fount, means the design of the printed characters.

Question 23.1

Why is colour printing of an image with an inkjet printer slower than text printing in black?

The selections of print quality are made by controlling the nozzle movement so as to alter the number of ink-dots per inch. In draft mode, modern inkjet printers use a resolution of 300×300 dots per inch (the same resolution both horizontally and vertically). In the higher quality modes, the resolution is normally 600×600 dots per inch. For graphics, the printer head drive and paper feed can be controlled so that dots overlap, so that resolutions that appear to be much higher can be obtained, typically up to 1700 dots per inch. The true resolution, however, is that set by the number of nozzles on the jet-head. Attempts to obtain very high apparent resolution require a combination of ink and paper that avoids one dot of ink spreading too much on the paper, and high-resolution pictures often come off the printer quite wet and need to be carefully dried to avoid smudging.

Controls

At one time, printers used control panels with a set of switches and, often, an LCD display. Modern printers are totally controlled by software in the computer, so that the controls on the printer itself nowadays amount to a few switches and LED indicators. Some types even dispense with an on/off switch and are set to a standby mode by the computer.

A typical modern printer will use two buttons and three LEDs. One button is used for power on/off, and the other to resume normal printing after an interruption (such as replacing paper or a cartridge). The power button may act only on a low-voltage DC supply if the printer takes its power from a mains adapter, and you might want to ensure that the power supply unit can be switched off at the mains when all printing is finished.

- Some printers are fussy about the order of switching down, and you are required to switch off at the printer before switching off at the mains.

One LED will be used to indicate power on, one to show that the *Resume* button needs to be pressed, and one to indicate that a cartridge needs to be replaced.

- Whatever system your printer uses, you should place a reminder about these controls close to the printer, so that a user does not need to find the manual in a hurry when LEDs are lit or flashing.

Cables

The cables that need to be connected to the inkjet printer are the power (mains) cable and the printer cable. Some printers contain a power supply internally and are connected directly to AC mains, more commonly now, the printer uses a low-voltage DC supply, and is powered through a transformer/rectifier supply that plugs directly into a power socket, or has an AC power cable attached. Typical power requirements are of the order of 5 W in standby and 30 W when printing.

The printer data cable needs more consideration. Older printers could use an ordinary parallel (Centronics) type of cable, but modern printers need the bi-directional type termed IEEE 1284. This type of cable allows two-way communication between the printer and the computer, so that the printer can signal to the computer when it is out of paper or ink. The IEEE 1284 cable also requires the computer to be set up so that its parallel port is of the ECP type – this usually has to be done using the CMOS RAM Setup action.

Paper

Paper for text work on inkjet machines can be any normal office paper marked as suitable for photocopying, laser or inkjet printers. The specification of paper thickness is in terms of the weight of a square metre of paper, and typical weights for inkjet printing are 75 to 90 g/m^2. The use of expensive coated papers is justified only if colour photographs are being reproduced. The best-quality thick photo paper can cost 50p or more per sheet. If you need to reproduce photographs on thick paper, make sure that your printer can handle such paper.

- Paper needs to be stored in a cool dry place. Before loading paper into an input tray, the sheets should be shuffled to ensure that they are not sticking together – if more than one sheet enters the printer it is likely to cause a jam.

Installation

A new printer will be delivered in a cardboard case, and this should be opened carefully to avoid damaging the contents or spilling packing materials. A typical pack will contain the printer, its power supply/mains cable, cartridges, software on a CD-ROM, and an instruction manual. These should be checked to ensure that you have all that you need. The printer cable is not usually included with the printer, giving you the choice of using a parallel IEEE 1284 type of cable or a USB cable.

- You should always keep all packing material and cases in the event of having to return or move the printer.

When the unpacking is complete and the items checked, the printer can be set up. Internal packing is normally used to prevent the printheads moving while the printer is being transported, and all of this packing must be removed before any attempt is made to connect up the printer. Other moving parts may be taped to prevent movement, and the tape also must be removed. The packing and taping will be described either in the printer manual or in a separate document. Keep the packing and the information in the event of having to repack the printer.

Once the packing has been removed, check that the printer is in good condition, with no apparent damage, and place it on a flat surface. This need not necessarily be where it will finally be used, but it's often more convenient to install the ink cartridge(s) when the printer is easily accessible. When you have to replace a cartridge, it will not matter if the printer is less accessible because you will already have gone through the procedure. The printer can now be connected to the computer and to the power supply for testing.

First, the ink cartridge(s) need to be inserted. Each cartridge comes in a pack and must be carefully removed. The ink nozzles on the cartridge are usually covered by a strip of transparent tape which has to be removed before the cartridge is inserted. Before removing the tape, check how the cartridge fits into the cradle, and where the retaining clip engages. When you are certain that you know how this is done, remove the tape, press the cartridge into place and clip it securely.

- Different machines use different methods of retaining cartridges, so that you will have to check with the supplied manual.

- Note that a colour printer will need to have its colour cartridge installed even if you never intend to print in colour. In general, machines that use more than one cartridge will not work if a cartridge is missing.

Some printers can be self-tested at this point. If so, load in some paper, and press the self-test button to see a sheet printed. It's more likely that a modern machine makes no provision for this action.

Driver installation

The next step is to install the Windows driver. This software converts the computer data into signals that the printer can use to produce text and graphics on the paper, and it's important to use a driver that is correct for your printer.

Driver installation starts with establishing the correct printer description. You need to know the name of the manufacturer and the precise model number of the printer. For example, if you are installing a Hewlett-Packard DeskJet 895Cxi, it is not good enough to use the driver for the 895 series – you must look for the precise model.

With everything connected up and switched on, run Windows and click on the Start button, then on Settings and Printers. Now click on the icon labelled *New Printer*. Double-click (or click, depending on which Windows version you are using) on this icon to start a Wizard which will install your new printer.

A pair of lists will appear, one on the left-hand side of printer manufacturers and another on the right-hand side of models corresponding to the manufacturer whose name you have clicked. Once you have clicked manufacturer and model, you can proceed to the next part of the installation which will ask you to insert the CD-ROM (or floppies) that you used to install Windows. The driver software will be read from this source, and you can opt to print a trial page to ensure that the driver software will operate your printer correctly.

A slightly different procedure is needed if your printer is one that is not listed. This is likely if the printer was not manufactured when your version of Windows was issued, and you then need to use the software supplied by the printer manufacturer (which you can ignore if you find the make and model listed in Windows Add Printer lists). In this case, you ignore the list, insert the floppy containing the drivers, and click the button marked *Have Disk*. You will then be guided by messages on the screen so that the driver is installed from the disk, and you will usually be asked to opt for printing a test page.

With the Windows driver installed, you can now print from any program that runs using Windows. This is one of the benefits of using Windows, because when you use the older command-line type of operating system (such as MS-DOS) you will need to use a separate printer driver for each program that you run.

Activity 23.1

Connect up a printer, and install the cartridge(s). Boot the computer, and install the printer driver. Print a test page. If alignment of a colour cartridge is needed, carry this out.

Routine maintenance

Modern inkjet printers require little maintenance other than cleaning and cartridge replacement. The exterior of the printer should be wiped at intervals with a soft moist cloth. You must not use any form of solvents or cleaning fluids, and you should not normally attempt to clean the interior of the printer unless a spillage of ink has occurred. You should not lubricate the rail(s) on which the cartridge cradle slides. The cartridge cradle may need to be cleaned (check with the manual), using a clean, preferably synthetic, cloth.

In particular, you are warned in the documentation that accompanies your printer, not to attempt any cleaning action on the printhead nozzles, as this is likely to result in blockage. If nozzles become blocked, then the printhead should be replaced. On Hewlett-Packard machines, each cartridge is combined with a printhead, so that it is very unusual to have a problem with blocked nozzles. On other machines that use a separate ink tank and printhead, the printhead usually has to be replaced after a stated number of tank replacements (often 5). If the area around the nozzle block is dirty, it can be cleaned with a synthetic cloth, taking great care not to wipe the nozzle area.

In an emergency, a printhead that is giving problems may have to be cleaned. The first area to clean is the set of electrical contacts, gently wiping these with a moist cloth that will not deposit strands on the contacts. Try out the head after cleaning the contacts. If this has no effect, you can try gently wiping the jet area with a synthetic cloth, and if nothing else works, the drastic solution can be to boil the head in distilled water – this applies only to a head that can be detached from its ink tank, and is a desperate measure. The preferred solution to all jet problems is to replace the printhead.

Cartridge replacement needs to be done when a cartridge is empty, usually indicated by faint or streaky printing. The printer manual will indicate the precise procedure for your printer, but the general method is as follows:

1. With the printer switched on, use the control buttons to move the cradle to the centre of its rail(s). On some Hewlett-Packard machines, this is automatically done when the top cover is opened.
2. Switch the printer off and unplug it.

3. Unclip the cartridge and lift it out. Lay the cartridge on a scrap piece of paper.
4. Unpack the new cartridge.
5. Remove the tape covering the nozzles of the new cartridge.
6. Insert the new cartridge and clip it into place.

Once a cartridge has been replaced, restore power to the printer and use the controls to return the cradle to its rest position.

Diagnostics and utilities

Select the maintenance task you wish to perform.

☐ Align the print cartridges.

☐ Clean the print cartridges.

☐ Print a test page.

☐ Test printer communication.

Software diagnostics can be used for printers that are connected through the IEEE 1284 cable, using the ECP parallel port. For an inkjet printer, the main error messages are likely to concern out of paper, paper jam, and empty cartridge situations. The utility programs typically allow for aligning cartridges where more than one cartridge is used, for cleaning nozzles, printing a test page, and checking that there is communication in each direction between the printer and the computer.

The nozzle cleaning action is a utility that is used on several types of inkjet printers. It works by applying pulses of ink to each nozzle in turn, aiming to clear any blockages by pressure of ink. The routine should not be used if the ink supply is low, because it uses considerably more ink than normal printing.

Question 23.2

What would you expect to cause:

(a) white streaks across each line of a text document, (b) black streaks across each line of a text document?

Activity 23.2

Using an inkjet printer, carry out the manufacturer's recommended cleaning actions. Replace the black cartridge and print a test page.

Laser printer

The laser printer has become the standard office printer on the grounds of high-speed printing and quiet operation. The principle is totally different from that of dot-matrix printers, and is much closer to the photocopier principle, based on xerography. The heart of a laser printer is a drum made from light-sensitive material. This drum is an insulator, so that it can be electrically charged, but the electric charge will leak away in places where the drum has been struck by light. The principle of the laser printer (and the photocopier) is to charge the drum completely and then make the drum conductive in selected parts by being struck by a laser beam. The beam is switched on or off and scanned across the drum as the drum rotates, all controlled by the pattern of signals held in the memory of the printer, and enough memory must be present to store information for a complete page. A schematic for a typical laser printer is shown following, Figure 23.1.

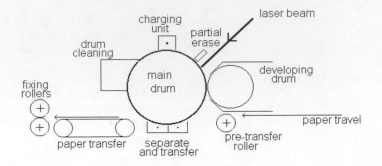

Figure 23.1   Outline of a laser printer mechanism

This system requires built-in memory of about 0.5 Mbyte as a minimum to store data for text work, and 2 Mbyte or more if elaborate high-resolution graphics patterns have to be printed. Once the scanning process is complete, the drum will contain on its surface an electrical voltage 'image' corresponding exactly to the pattern that exists in the memory, which in turn corresponds to the pattern of black dots that will make up the image. Finely powdered resin, the **toner**, will now be coated over the drum and will stick to it only where the electric charge is large – at each black dot of the original image.

The coating process is done by using another roller, the developing cylinder, which is in contact with the toner powder, a form of dry ink. The toner is a light dry powder which is a non-conductor and a scraper blade ensures that the coating is even. As the developing cylinder rolls close to the main drum, toner will be attracted across where the drum is electrically charged.

Rolling a sheet of paper over the drum will now pass the toner to the paper, using a corona discharge to attract the toner particles to the paper by placing a positive charge on to the paper. After the toner has been transferred, the charge on the paper has to be neutralized to prevent the paper from remaining wrapped round the drum, and this is done by the static-eliminator blade. This leaves the toner only very loosely adhering to the paper, and it needs to be fixed permanently into place by passing the paper between hot rollers which melt the toner into the paper, giving the glossy appearance that is the mark of a good laser printer. The drum is then cleared of any residual toner by a sweeping blade, re-charged and made ready for the next page.

The main consumables of this process are the toner and the drum. The toner for most modern copiers is contained in a replaceable cartridge, avoiding the need to decant this very fine powder from one container to another. The resin is comparatively harmless, but all fine powders are a risk to the lungs and also a risk of explosion. Drum replacement will, on average, be needed after each 80 000 copies, and less major maintenance after every 20 000 copies. Paper costs can be low because any paper that is suitable for copier use can be used – there is little advantage in using expensive paper, and some heavy-grade paper may cause problems of sticking in the rollers.

- As for an inkjet printer, the correct Windows driver must be installed in order to operate a laser printer satisfactorily.

- Laser printers for office use are generally black-ink printers, and colour laser printers are expensive and need very expensive cartridges.

- Empty cartridges should be returned to the manufacturer for recycling, or they can be donated to any charity shop that will accept them. Do no attempt to open or refill cartridges for yourself – toner dust is dangerous to the lungs if inhaled.

Printer comparisons

The cheapest monochrome laser printers for home use can be sold at much the same price as a colour inkjet model, but if we compare like for like, laser printers are more expensive to buy. For black-ink uses, the laser printer provides better-looking text and line drawings, but the difference between a laser print and a print from a good inkjet machine can be seen only with the aid of a magnifying glass. The laser lines look smoother, with no trace of separate dots or of smears. For colour reproduction, the laser copy is sharper and brighter, but if high-quality paper is used along with a photo-quality inkjet, the differences are small.

Printers of any type that are specified for office use are built to standards that assume almost continuous use at fast printing speeds, and are priced accordingly.

Inkjet printers can cost more in running costs, per printed page, due mainly to the cost of cartridges. The life of a laser cartridge (costing typically £40) is much longer than that of a typical inkjet cartridge at £25. If expensive inkjet paper is used, the difference in running costs becomes greater.

In general, laser printers are faster, though the claimed speeds can be very misleading. The claimed speed for a laser printer is usually for making a large number of copies of a single page, and printing speed for different pages is lower, because of the time needed to create a different pattern on the drum for each page. Claimed speeds range from 6 pages per minute to more than 20 pages per minute in monochrome. The fastest laser printers can turn out monochrome documents at around twice the rate of the fastest inkjet machines.

Answers to questions

23.1   The carriage must be moved to print each coloured dot in a line, and the roller must also be moved for each coloured dot vertically, making overall printing speed much slower.

23.2   (a) A jet is blocked.

(b) A paper fibre, soaked in ink, is attached to the printhead.

Appendix 1

Certification and assessment structure

The NVQs/RVQs cover the general core studies for radio, audio, television, industrial electrical/electronic systems, refrigeration, domestic cookers and gas and other small appliances. The syllabuses are designed for the vocational education of new entrants to the servicing industry. They specifically meet the requirements for the under-pinning technical knowledge that is needed to obtain the appropriate award. The major difference between the two awards NVQ and RVQ which are considered to be off equal status, lies in the way in which candidates progress is accessed and accredited. For the NVQ, the college or training organization, assesses both the under-pinning technical knowledge and practical competence. The overall progress may be moderated and accredited by a training authority such as EMTA. For the RVQ which is progressively replacing the C&G 2240 and EEB structures, the changes will only be marginal. The C&G will still examine, by multiple choice question papers, the under-pinning technical knowledge, whilst the assessment of the practical competence topics will be the joint responsibility of the training organization and the EEB.

At all times, the health and safety aspect within both the work place and the area in which the systems are employed is specifically emphasized.

The Level 2 qualifications include five mandatory core units plus one optional unit. These candidates are expected to be able to work to clearly defined procedures and with responsibility for identifying and implementing any decisions. However, they will need to refer to other competent servicing personnel for authorization and guidance. Fault finding and repair for this award is restricted to module or stage level.

The Level 3 qualification requires a thorough understanding of four mandatory core modules plus two optional units. The candidates must demonstrate an ability to work to procedures that are not clearly defined. They must take responsibility for planning and controlling the quality of work, and to instruct and give guidance to other less well qualified servicing staff. Level 3 candidates must be able to demonstrate their ability to fault find and carry out repairs down to component level.

In addition to all this, all candidates must be able to demonstrate an ability to carry out soldering and de-soldering exercises and test for the serviceability of components.

Appendix 2

General syllabus structure for NVQ Level 2

The registration and certification for this course structure is a continuing process, with candidates' progress being monitored and tested and logged at each stage by appropriately qualified personnel. The Contents listing at the beginning of this book, reflects fairly accurately, the topic headings covered by the syllabus for the common core technology needed for the Level 2 award.

For all disciplines of this new course, all candidates must have a good understanding of the Health & Safety Regulations and the associated legislation as it affects the servicing industry. This aspect which affects both the workplace and the areas in which the equipment is employed, means that many of the specific applications can only be certified within the normal working environment.

Technically, each candidate is expected to demonstrate the ability to carry out testing and diagnosis down to stage or module level, including the correct dismantling and re-assembly. This includes the ability to recognize that the test equipment being used is operating correctly and is calibrated. After repair, the candidate must be able to re-install any equipment back in its original place of use to the satisfaction of the owner or user.

Particularly in an industrial environment, candidates are expected to be able to follow the instructions of any planned maintenance scheme and carry out the necessary updating of its database.

Appendix 3

Multiple-choice examination questions: guidance for examination candidates

The certifying and examining bodies for electronics/electrical servicing rely considerably on the use of multiple-choice questions at all levels. The sheer size of a multiple-choice question paper often leads candidates into feeling that time is short, leading to a feeling of urgency which results in bad or even silly choices being made. The first point to understand, then is that the time allocation is in fact quite generous, and the size of the paper is due to the space that is needed for printing diagrams and for the choice of answers.

The golden rule for answering this type of questioning is to formulate your own answer to the question before looking at the choice of answers, and then selecting the answer that most closely fits with your own. In this way, when you know the answer as you read the question you will not be distracted by reading over the list of choices, all but one of which will be designed to distract you. On your first reading of a paper, answer all of the questions that you can using this method, and only then proceed to look at the others.

If you are unsure of an answer of your own, then you can usually reject all but two of the possible answers that have been provided. If you are genuinely uncertain about the correct choice of these two, then you can opt for one or other knowing that your chances of being correct are considerably higher, 50 per cent instead of 25 per cent than they would have been if you had simply guessed in the first place.

A methodical approach like this can considerably improve any candidate's score in a multiple-choice examination, particularly when there are usually two or more possible distracting solutions that are most likely to be wrong.

INDEX